Journey to Clarity

Journey to Clarity

David A. Tozar

Copyright © 2017 by David A. Tozar.

Library of Congress Control Number:		2017909626
ISBN:	Hardcover	978-1-5434-3102-5
	Softcover	978-1-5434-3101-8
	eBook	978-1-5434-3100-1

All rights reserved. No part of this book may be reproduced or transmitted in any form or by any means, electronic or mechanical, including photocopying, recording, or by any information storage and retrieval system, without permission in writing from the copyright owner.

Scripture quotations marked NKJV are taken from the New King James Version. Copyright © 1982 by Thomas Nelson, Inc. Used by permission. All rights reserved.

Any people depicted in stock imagery provided by Thinkstock are models, and such images are being used for illustrative purposes only.
Certain stock imagery © Thinkstock.

Print information available on the last page.

Rev. date: 07/05/2017

To order additional copies of this book, contact:
Xlibris
1-888-795-4274
www.Xlibris.com
Orders@Xlibris.com
762592

Contents

Chapter One: I Am Born 1
Chapter Two: World Views 11
Chapter Three: The Arrival of the Universe 42
Chapter Four: It's Alive! 64
Chapter Five: My, How You've Changed 77
Chapter Six: Intermission 93
Chapter Seven: Crossing the Line 98
Chapter Eight: New Testament Records 130
Chapter Nine: Stating the Case 149
Chapter Ten: The Founding of the Church 175
Chapter Eleven: Q&A 215
Epilogue 239
Acknowledgments 247
Sources and Suggested Reading 249
Appendix A 251

To my wife, Jean, the angel on my shoulder, my earthly rock, and the love of my life. And to my sons, Gabriel and Elijah and grandson Tanner.

Here I stand.

Chapter One

I AM BORN

I think it necessary before departing on our journey to give you a little context regarding my frame of mind when I initially started my solo trek so many years ago. Now, I know that we are all unique and our individual starting points will be many and varied. However, we all do possess some common characteristics outside of those that are specifically animal in nature. Beyond our need to breathe, nourish, and such like, we are all also endowed with the ability to think and reason. This does not mean that we all use these attributes to their full advantage as we should (as evidenced by some people that I'm sure you've met along your respective roads), but we all have the *ability*. Indeed, I wish that I had thought some things through more completely, been more incisive, than I did because I may have relieved myself of a great deal of pain and embarrassment. This too we can count as a common characteristic if we are honest with ourselves. Those people we have considered to be less intelligent than ourselves could well have been us at other times and places and probably were.

We are also the result of a string of people who came before us that we refer to as our ancestors. Some of yours may have been sheep farmers, weavers, peasants, noblemen, or cattle rustlers hanged for their choice of vocation. For better or worse, our composition, at least physically, is a conglomeration of those chromosomes. I think there must have been a couple of snakes in the woodpile in my past based on the way I turned out. After all, being the way we are, I have got to blame it on something other than some of the choices I've made. This too we hold in common.

It is with this family environment that we are most concerned. Our family and the friends we made along the way determined much of what we grew up believing. This group, including teachers and perhaps others based on our respective situations, has made indelible impressions on us that are difficult to correct if those impressions were founded on error. These folks have passed on to us, whether with good intentions or otherwise, what they have learned; and if significant thought or questioning had not been part of the equation, we may have ingested much rotten fruit. It is with the knowledge that the world is full of rotten fruit, fed to us with both malicious and benign intent, that makes our journey necessary.

So the world of my youth is revealed to you, at least partially. You will be able to substitute yours in its place to determine your starting point. I do believe that a certain amount of introspection is necessary, as well as sober mindedness, for us to evaluate the steps we take to clarity. You must have an open mind and be brave enough (there will be those you say naive enough) to follow the trail wherever it might lead.

One Side of the House

A family, or at least some of the more adventurous men, surnamed Berold crossed the English Channel with William, Duke of Normandy, in 1066 with conquest in their minds and booty in their hearts. After the conquest, as the centuries washed over them, they were assimilated into the culture they had so rudely interrupted, and their name was Anglicized into Barrett or, sometimes, Barratt. In the 1640s and 1650s Barretts settled in the New World around Cape May in New Jersey. I don't know that their removal thereto was because of anything as romantic as religious persecution. It may be that they still had a smoldering desire for conquest and booty, but remove there they did. During this same period, some Skulls and Steelmans also crossed the great Atlantic, seeking a new beginning.

The Steelman contingent may have been following a fellow Swede, a sea captain named Hans Mansson, a.k.a. Monsieur a.k.a. Steelman, who had crossed and stayed in 1642. I don't know what his crew thought of his deciding to stay. I have never spoken to any of them. They're dead, as is Hans. In any event, Hans liked the looks of New Sweden (today's Delaware; I'm sure it was much nicer then), and that is where he settled. In 1654, he married a woman named Ella.

The reason that I have mentioned all this is because these folks represent the beginning of my family in America and, because of guilt by association, the beginning of me. These Barretts, Skulls (who changed the spelling of their name to Scull; I would have too) and Steelmans must have liked one another because there was a good deal of intermarrying going on among them. Or it may be that their choices were limited as there were only around 130 people (including the Barretts, Sculls, and Steelmans) inhabiting Cape May at that time. Either way, it all worked out.

When the American Revolution came along, one Elijah Barrett was enlisted as an ensign (kind of a sublieutenant) into the Seventh Company, Third Regiment of the Gloucester County Militia. My youngest son is his namesake. Service in the United States Military would become more or less a family tradition. It is this Barrett that gained my admission into the Sons of the American Revolution though I am not a member of the snooty sect.

In 1740, Captain George May sailed up the Great Egg Harbor River from Cape May and purchased land in 1756, which would become the town of Mays Landing. Prior to Captain May, the residents of this area were largely Absegami Indians who were a segment of the Lenni Lenape tribe. Around 1695, settlers were recorded as having moved into the area; and by the early 1700s, Methodist and Presbyterian missionaries were making trips into the interior. In 1710, a distant relative, Peter Steelman, first settled in what was then called Ilifftown because of the rather large landholdings of one Edmond Illiff. When Captain May sailed up the Great Egg Harbor, he did so at the behest of the London Company, and he found a land rich in oak and pine perfect for shipbuilding. This was the land he purchased, and within a short time, he had established a trading post and shipyard. By 1778 the center of the town was recognized by the inhabitants as Mays Landing.

It is this town and Cape May that I remember as a child. My dad's family were here and had been for a couple of centuries. My grandfather's house backed up to the Great Egg Harbor River, and I spent a lot of the time on that river when we were there visiting. I will always remember that river as perfectly black, slow moving, and oily. The banks were crowded closely by the woods and undergrowth. Turtles came out to sun on the protruding trunks of trees that had fallen into the water because their roots had gotten too close to the edge. (My brother and I had been warned about getting too close. I guess those trees didn't have a mama.)

I thought at the time that Pop Pop's (as we called him) house must have been built when the first settlers entered the area. It certainly looked like it. The kitchen had a pump handle to provide the only water to the house, and a "three-holer" stood out back. There were times in the summer that I thought it stood a tad too close. No one had even thought of painting or repairing this dark edifice for one hundred years, at least. Pop Pop and Mom Mom were too poor to even think about such things.

The house, as I later found out, was built in 1796 though I'm not sure who the original builder was. It consisted of three stories, and my dad had the lone bedroom that was in the peak of the house. There was a stove in the kitchen that provided the only heat source for all three floors and was also used for cooking. When my dad was a boy, it burned wood but was later converted to kerosene. (If I close my eyes, I can still see the wallpaper peeling off the walls and smell the kerosene that permeated the entire structure.) When he awoke on winter mornings, he had to break the ice out of the pitcher on his washstand to be able to wash himself. In spite of this, Dad seemed to have a typical childhood though there certainly were periods of want. Many dinners consisted of whatever he or his father managed to catch or trap. I believe my father didn't care much for this situation, but my grandfather loved it. He was an outdoorsman, and I believe that I have him to thank for that trait, which showed up in me.

My father did well in school and excelled when he entered high school. There was, however, no question of a college education; so upon graduation, in 1938, he enlisted. Frankly, anything was better to him than staying in Mays Landing. The army would allow him to travel; it would provide him with three meals a day that he had neither to catch nor clean and give him a bed in a room with heat—sometimes. And on top of all this luxury, they would pay him! What could be better? Aside from trips to visit his relatives while he and they lived, he never went back to Mays Landing. I believe he was the only one of my direct relatives who ever left that town. They were an intensely strange lot to a Southern boy (me) who had grown up in Virginia, though always very kind to my mother, brother, and me. I believe that back then everyone that had been in Mays Landing for any length of time was related in some fashion.

Dad was enlisted into the Sixteenth Regiment of the First Infantry Division, which was being formed up at Fort Jay in New York. As I mentioned previously, the year was 1938, and little thought of war was in anyone's mind (at least in the United States). I think that the public schools still teach that the Japanese attacked Pearl Harbor on December 7, 1941, but I'm not sure. Either way, they did, and many American men and women became embroiled in the conflict whether by choice or coercion. Nearly every family in our country was involved in some facet of the war effort; men in the service if fit for it, women in the factories and the various service auxiliary corps, children in scrap drives. America staggered the world with its ability to produce tanks, planes, small arms, long guns, and ships as well as uniforms, belt buckles, knapsacks, boots, and socks. We were totally committed to ending Imperial Japan and Nazi Germany, period.

My father was a combat medic. He went ashore in North Africa for Operation Torch in 1942. After the Americans and British (with some assistance from the French) completed this job, their next target was Sicily. Here too he got his boots and fatigues wet and continued on until the island was liberated and the remaining German troops surrendered or departed for Italy. From Sicily, he went to England to prepare for the Normandy invasion. The First Infantry Division was assigned Omaha Beach along with the Twenty-Ninth; and at 6:30 a.m. on June 6, 1944, Dad went in with his regiment.

During this period, combat medics wore a red cross on a white armband for identification as, at least in theory, this would provide them with some protection while they went about their duties caring for the wounded. Many German soldiers either could not or would not separate them from other targets of opportunity, and their casualties were proportionate with the riflemen. From Normandy, on across France, the Battle of the Bulge, and into Germany he trod along with tens of thousands of others leaving their dead behind to be policed up and buried by those assigned to that duty. When it was all over, he returned home, remained in the army, and went to work in the Pentagon.

The Other Side

Around 1700, a gaggle of Scots came to America, and many of them settled in North Carolina. From that influx of Scotsmen, my mother was descended. Her family was made up of Stricklands, Dawes, and Joyners. There were probably others, but General William T. Sherman and his army made sure that I couldn't find out about them. In 1864–65, he marched from Atlanta, Georgia, to the coast of South Carolina, burning everything that would burn and destroying what wouldn't. When he left the coast, he turned northwestward with the intention of linking up with his benefactor and superior officer Ulysses S. Grant around Richmond or wherever Grant had managed to push General Lee in the meantime. Sherman was to chase Gen. Joseph E. Johnston before him if he could not be destroyed. The area of North Carolina my ancestors lived in lay right in his path, and that area suffered the same fate as that visited upon the line of march farther to the west and south. All the family records earlier than those retained in my grandfather's Bible went up with the smoke and ashes of the Union Army's fires.

My grandfather, Russell Strickland, arrived in the small town of Elm City in the late 1890s and married Ms. Sudie Joyner shortly thereafter. He was the postmaster for many years in this dusty, sleepy town. In 1905 my uncle, Alton, was born and my mother, Ruth, in 1911. Following the end of World War I, the great influenza pandemic was brought to America, and in 1919 it took my grandmother along with millions of others worldwide. My uncle was old enough to stay with his father, but Mother was sent to live with Grandfather's sister, Lizzie (Elizabeth).

I probably don't need to go into much detail to convince you that growing up in the rural South at that time was considerably different than childhood in New Jersey. Heck, it's still different. They may well be considered opposite poles on a completely different planet. The town of my mother was full of horse drawn wagons, farmers, and aging Confederate veterans hanging about outside the drug or hardware stores. The Elm City I remember was of quiet, lightly traveled streets without lighting or sidewalks. The magnolia and oak trees were old and grew together to form an archway over them. As a child visiting there, I largely went barefoot unless some strenuous activity required shoes. One oddity that I remember from those days in Elm City (and the surrounding towns where other relatives lived) was the party line. You had to clear off someone else who may be on the phone line before your call could be connected.

Of course, Sunday was a special, holy day not to be interrupted by the shouts of children playing or by otherwise behaving indecorously. The Baptist church was wood framed, hot, and stuffy. Despite this, men who had them wore suits and women their dresses and bonnets. Some men, straight off the farm, wore bib overalls. The pews also held fans for the ladies and Baptist hymnals but were void of Bibles because everyone had their own and carried them unfailingly in hand or under their arm to leaving one hand free to shake hands which they did all around. The services were pious and solemn, and there was much gossiping after they ended. I can't say I remember much about them as I always struggled to stay awake. There was no "children's church" as exists today.

In 1928, my mother left her hometown to attend East Carolina Teachers College (now East Carolina University) to get her teaching certificate. It was a two-year school at the time, and having completed her course successfully, a professor talked to her about continuing on to get her four-year degree. While Mom was willing, the family finances would not allow it. Fortunately, at least for her future, an uncle died at that time and left her the whopping sum of $350. This afforded her the opportunity of enrolling at Atlantic Christian College in Wilson, North Carolina (now Barton College). When she graduated in 1932, she only owed the school $45. Being diligent in her coursework, she was allowed part-time employment as a tutor and assistant librarian, which assisted her with books and board. After graduation, back to Elm City she went and embarked on a teaching career until, like all Americans, her life was interrupted by World War II.

Feeling that she needed to do whatever she could to assist in the war effort, she enlisted in the Women's Army Corps (WACs) as a nurse. She served in army hospitals in the States, nursing returning GI's back to health or at least that degree of health they were capable of so that they would be fit for a return to civilian life. Most cases that came all the way back to the United States would prove to be unable to return to duty. Some men never left the hospital, not alive anyway.

I believe it was sometime in 1944 that she was assigned to Walter Reed Army Hospital in Northern Virginia. As mentioned, Dad had come to the Pentagon at the wars' end, and it was on a blind date that Mom and Dad met. My father had a buddy named Chuck who was dating a nurse named Betty who happened to work with my mother in the hospital. Chuck invited Dad, Betty invited Mom, and the rest, as they say, is history. The attraction between Arthur Tozar and Ruth Strickland is difficult to figure, at least in my mind, they were married in 1949, departed the US Army, and settled in Northern Virginia. They loved each other beyond doubt and continued to do so until my father passed away in 1997. Mom lived until age ninety-two and left us in 2003. I'm sure that there must have been some clashes between them, but I can honestly say that I never heard either of them exchange a harsh word within my hearing.

This strange amalgamation of disparate personalities produced my brother, Russell, and me early in the 1950s. Things seemed a lot simpler then though the world was learning to go about its business under a cloud which threatened nuclear war. The public school I attended conducted periodic air-raid drills as, I expect, all the other schools in the country did. Chairman Mao had completed the conquest of China and written his Little Red Book. In Russia, Joseph Stalin had died and Nikita Khrushchev was now in control. The Cold War was ramping up and would haunt us for thirty years. The Cuban missile crisis and the Bay of Pigs were followed by the Vietnam War. John Kennedy was murdered before he could withdraw our troops, as we were told after the fact that he intended to do. Close on the heels of all this or—perhaps, better said—intermingled with it was the beginning of rock 'n' roll, the Beatles, Haight Ashbury, LSD, and a cultural revolution within our borders and throughout much of

Europe. I was left to try to become a responsible adult while all this was going on.

Now, it is only fair to say that many of my classmates in high school seemed to be perfectly normal, focused kids. Not so myself. I hadn't a clue what I was about. I'm not blaming my parents, but they had given me an unusual amount of leeway in selecting how I would spend my time. While still very young, I had discovered books. My father was enamored with them, and of course my mother was a teacher, so she also encouraged my interest. I read well beyond my age and probably read much that I had no business getting into. In fact, I'm sure that I did. When my father questioned my mother about this, I remember her response. She said, "Arthur, if he's able to understand what he's reading, let him read it." While my mother had the loveliest soul of anyone I've ever known, I think my father's instincts were more accurate on this particular account.

When I wasn't playing baseball, reading absorbed me. I read history, biography, Mao's Little Red Book, and much of the cultural literature being ginned out by an equally wandering, youthful America. I particularly enjoyed the Russian authors. My tendency was to rebel against tradition though there was nothing to rebel against that I could reasonably articulate. Soon I became bored with high school. Aside from the sciences and higher math, which held no interest for me, I had read beyond anything being taught in social studies or in the literature classes being offered. With only three months left before graduation, I left. After completing my G.E.D., I split.

The tumult that I was experiencing did not lead to arguments or any type of disrespect for my parents. Whatever was causing my unrest, I knew that it wasn't them. But at the same time, my parents never offered me any counsel. In their defense, by this time I was probably beyond being willing to listen to anyone. We never had discussions about religion though we went to church every Wednesday night and Sunday morning, at least until I was fourteen and was given the choice of going or not. I chose not. Mom and Dad never discussed the war years, their past, politics, or anything of importance with me. Quite frankly, I don't think they knew how to direct me. Mom was born in 1911 in the rural South. Dad was born in 1920 in quirky South Jersey. I believe that in many ways they were as befuddled as I was. My way was to be my own, and with my head full of disconnected facts, forming a purpose was beyond me. Such was my starting point on a journey that would take me years just to realize that I didn't even know what direction I was heading.

Chapter Two

WORLD VIEWS

Our journey will begin with establishing an understanding of what a world view is, how it impacts our everyday lives (even unconsciously), and what options for world views are available to us. To come to clarity, we will need to explore, at least to an extent, several topics that may be unfamiliar to us. They are, however, part and parcel of the package that makes up that by which we, as individuals, make sense of our experiences—in short, the way we view the world around us and our place in it. Since this is a labor of love, I will not subject you to all the trials, errors, and blind alleys I dealt with because of undisciplined and erratic study but will rather strive to lay out my findings in a coherent, rational order.

Why is this idea of a world view significant? Specifically, why should I care? Before we begin to answer these question corporately, allow me to posit a rhetorical one as kind of an ice breaker. Is the world messed up? If so, how did it get this way? Okay, that's two, so while I'm at it, I'll toss out a couple of others. What can we do to fix this messed-up situation? *Can* we fix it? Where did we come from? Since we all live here and must deal with the situation as it is, I would submit that we all have a vested interest in the answers to these questions. How we view the world and our experiences in it directly influences the activities we either support or seek to stop. It will lead to our respective ideas of what a "fixed" planet to live in looks like.

Look at the situation as it stands today—2017. It seems that there are more and varied interests vying for the right to direct us to the promised land than ever before in recorded history. In our country, there are liberals who want us to be peaceful, tolerant, and loving though they demonstrate none of those traits themselves when you happen to disagree with them. Of course, on the other end of the spectrum, we have the conservatives though there are so few of them that are truly conservative that we can hardly tell who they are or how conservatism should be defined. We also have a smattering of socialists/communists who, in my opinion, are so limited in their vision that they can't see the contradictions inherent in their system clearly enough to recognize the silliness of their position.

In the world at large, we have militant Muslims (a minority, I believe) who want to convert or kill, various communist or despotic regimes who likewise want direction of our futures come what may, and many others who are unheard of because they haven't yet gotten the airtime. The way in which we choose to live our individual lives and the world we want to leave to our children is the direct outgrowth of the seeds we ourselves plant. These seeds drop from the tree of our world view.

Let's Begin

From the records we have available to us today, every civilization that we know of has attempted to understand how the world around them worked, the means by which they came to be, and the power or forces that caused it. Predicated on their understanding at the time, they generally chose to create deities that controlled everything from the rising of the sun, annual rainfall, fertility, the outcome of war, the success (or not) of the harvest, and practically anything else you can think of. The ancient Greeks, for example, so beautifully made familiar to us by Homer and others, had their panoply of gods and goddesses. High on Mount Olympus, you could hardly swing a cat without hitting deity.

Reaching even further into the strange and exotic past, we find the Sumerians, Mesopotamians, Babylonians, Egyptians, and many others attempting to come to grips with this beautiful yet tenuous thing we call life. Some had deeply ingrained belief systems; others clearly derived theirs from either borrowing certain aspects from neighbors or co-opting the entire regime of deities and simply renaming them. Much like, in later days, the Romans did the Greeks.

About 400 BC, however, a seismic shift in perception, or at least in the explanations associated with those perceptions, occurred in and around Athens, Greece. Why Greece? Perhaps it was the warm Mediterranean breezes, I don't know, and I'm not sure anyone else in modern times does either. Happen, however, it did; and the ideas surrounding man, the world, and man's place in it changed forever.

For myself, I chose the murky depths of philosophy as the beginning of my study. Unfortunately, I was lost in the weeds for longer than I care to admit. Dark indeed where those times with much more of depression than delight. Looking back, I realize now that it was not the subject but the teachers who got me lost. (Yes, really.) You see, I started with modern philosophers thinking that they would have the most up-to-date thoughts on things. My mistake. It now has become plain to me that most of what was worthwhile to be said of a philosophical nature was said a couple of thousand years ago. I did not find any clarity contained in the writings of the moderns (with a couple of important exceptions).

It has been said that all men (and women) are philosophers. This is usually translated to mean "You have an opinion. Therefore, you are a philosopher." You may subscribe to this particular idea, but it will not hold water in our journey. Our definition of philosophy will be "the study of knowledge, truth, and the nature and meaning of life so that one is thereby able develop a rational set of ideas about how one ought to live." Defined as such, the Greeks developed, and spawned, a number schools to develop and disseminate their ideas. Believe me when I tell you that there were a *lot* of folks who called themselves philosophers. Like their many gods and goddesses residing on Olympus, it seems you couldn't toss a dog in any direction without taking the chance of knocking over someone teaching philosophy or studying it. The schools at the time would also include mathematics, rhetoric and oratory, law (such as it was), and lesser topics that we, today, do not associate directly with the subject of philosophy. While there were indeed many philosophers, happily for us, there are two who will adequately serve as representatives for the opposing views that are important to our study.

Aristotle

Son of a physician named Nicomachus, Aristotle was born in the colony of Stagira in Thrace. The geography of ancient Thrace is now contained in southeastern Bulgaria, northeastern Greece, and European Turkey. The largest portion is in Bulgaria. The Thracians were first referred to in Homer's *Iliad* as allies of the Trojans. Settling initially as individual tribes, they had no particular clout until they formed into the Ordrysian State in the fifth-century BC. Being mountain folk, they were viewed by the flatland Greeks as fierce, warlike people. (Apparently, the Greeks considered themselves passive, lamblike creatures though I'm not sure history bears them out in this.)

Nicomachus was so well respected in his profession that he was offered the post of physician to Amyntas, king of Macedonia, which, lucrative gig that it was, he accepted. Amyntas was father to Phillip who would one day be king in his own right. Nicomachus was a man of many interests who wrote several books on natural history. We may safely infer Aristotle's future talents for acute observation and attention to detail were inherited from his father (no slight to Mom intended).

Aristotle

Born in 384 BC, his father died when he was eighteen years of age whereupon he repaired to Athens to study at the Academy under the master Plato. Here he would remain for twenty years until Plato, in his turn, departed this life. Aristotle's tenure under Plato was probably at times tempestuous as demonstrated by the fact that when Plato died, he bequeathed his school to a disciple named Speusippus and not to his famous understudy, Aristotle. This gentleman in no wise possessed the quality of mind contained in the skull of Aristotle. Many explanations for this removal from the will (so to speak) have been offered, but in all probability, it was simply a matter of the understudy not having accepted all of the great Plato's teaching without argument. Genius seldom does. In fairness, it should be noted that as a non-Athenian, Aristotle could not own the property that the Academy stood on so that may have entered into it as well.

In any event, Aristotle took himself off to a place named Atarneus (located in modern Turkey near Dikili in Izmir Province; it is on the mainland, just opposite the island of Lesbos) where he married the niece—or daughter, we're not sure which—of the local tyrant Hermias. His bride was named Pythias the Elder and was something of a prodigy in her own right in the study of biology. She was called the Elder not to be unkind but because she gave birth to Aristotle's daughter, also name Pythias. (You guessed it, called the Younger.) As a youth, Hermias had studied at the academy under Plato where he first met his future son-in-law. After Plato's death (347 BC), Aristotle and his buddy Xenocrates traveled to Assos under the patronage of Hermias.

Hermias's territory was allied with those who revolted against the Persian king Artaxerxes III, who was also the pharaoh of Egypt, and this king sent Memnon of Rhodes to coerce them back into line. By subterfuge, Memnon coaxed Hermias to attend a council to discuss the situation whereupon, directly upon his arrival, he was taken into custody and, in chains, sent to Susa for "interrogation" presumably to learn more about the intentions of Philip of Macedonia, leader of the insurrection. He and his bride continued to reside there for three years.

Despite the passage of twenty-three years since he was last in Macedonia, Philip (now king) had not forgotten Aristotle. He invited him to return to Macedonia and take up the instruction of his young son, Alexander. Serving in this capacity, he remained with the boy for eight years. Such a bond was formed between teacher and student that Alexander would send home specimens of animal life (while taking a break from the hard work of conquest) to assist in the master's proposed *History of Animals*. Indeed, at one point Alexander awarded Aristotle an enormous sum of money which allowed him to continue his studies without being molested by the gloomy aspects of want. When Philip died, Aristotle returned to Athens and opened his own school in the Lyceum. (Alexander, of course, went on to conquer much of the known world and to spreading Greek culture and joy along the way. Folks were so gratified by the Hellenization, for so it was called, of their lands that they forever after referred to Alexander as "the Great.")

As undoubtedly interesting as Aristotle's life was, it is, of course, with his writings—his mind—that we are primarily concerned. While perhaps a disruptive student at times, he was certainly listening to Plato as their philosophies coincide in many ways. Where the pupil completely breaks with his mentor, however, is in the technical construction of his ideas, proofs, and conclusions. All the fanciful allegory of Plato is carved away in remorseless fashion. In fact, so complete, so exhaustive down to the finest details was his philosophy that he left little for the schools to teach after he died but himself. This was true not only in Greece. In every part of the world that Greece had touched, he reigned supreme.

The strength and beauty of Aristotle lies in the fact that he was able to connect all the strings of that which is knowable about a thought or idea into a single scheme. He also was incredibly skilled in illustrating every detail of his whole. Or again, he tears the greatest concept down into its tiniest component parts, or he could take the tiniest part and build from it a majestic, sweeping concept. As he says in his *Metaphysics*,

> Wonder is and always has been the first incentive to philosophy. At first men wondered what puzzled them near at hand, then by gradual advance they came to notice and wonder at things still greater, as at the phases of the moon, the eclipses of sun and moon, the wonders of the stars, and the origin of the universe. Now he who is puzzled and in a maze, regards himself as a know-nothing; wherefore the philosopher is apt to be fond of wondrous tales or myths. And inasmuch as it was a consciousness of ignorance that drove men to philosophy, it is for the correction of this ignorance, and not for any material utility, that the pursuit of knowledge exists. Indeed it is, as a rule, only

when all other wants are well supplied that, by way of ease and recreation, men turn to this inquiry. And thus, since no satisfaction beyond itself is sought by philosophy, we speak of it as we speak of free men. We call that man free whose existence is for himself and not for another; so also philosophy is of all the sciences the only one that is free, for it alone exists for itself.

Moreover, this philosophy, which is the investigation of the *first causes* (emphasis added) of things, is the most truly educative among the sciences. For instructors are persons who show us the causes of things. And knowledge for the sake of knowledge belongs most properly to that inquiry which deals with what is most truly a matter of knowledge. For he who is seeking knowledge for its own sake will choose to have that knowledge which most truly deserves the name, the knowledge, namely, of what most truly appertains to knowledge. Now the things that most truly appertain to knowledge are the first causes; for in virtue of one's possession of these, and by deduction from these, all else comes to be known; we do not come to know them through what is inferior to them and underlying them. . . . The wise man ought, therefore, to know not only those things which are the outcome and product of first causes, he must be possessed of the truth as to the first causes themselves. And wisdom indeed is just this thoughtful science, a science of what is highest, not truncated of its head.

Elsewhere: "To the man, therefore, who has in fullest measure this knowledge of universals, all knowledge must lie to hand; for in a way he knows all that underlies them. Yet in a sense these universals are what men find hardest to apprehend, because they stand at the furthest extremity for the perceptions of sense."

Further, "Yet if anything exist which is eternal, immovable, freed from gross matter, the contemplative science alone can apprehend this. Physical science certainly cannot, for physics is of that which is ever in flux; nor can mathematical science apprehend it; we must look to a mode of science prior to and higher than both. The objects of physics are neither unchangeable nor free from matter; the objects of mathematics are indeed unchangeable, but we can hardly say they are free from matter; they have certainly relations with matter. But the first and highest science has to do with that which is unmoved and apart from matter; its function is with the eternal first causes of things. There are therefore three modes of theoretical inquiry: the science of physics, the science of mathematics, the science of God. For if the divine is anywhere, it must be in that form of existence I have spoken of (i.e. in first causes). . . . If, therefore there be any form of existence immovable, this we must regard as prior, and the philosophy of this we must consider the first philosophy, universal for the same reason that it is first. It deals with existence as such, inquiring what is and what are is attributes as pure existence."

I have included the quotes above as they have a direct influence on our further inquiry. Aristotle wrote extensively and minutely to develop his arguments, but as this is not a philosophical treatise, I will not attempt to relate the details in full that allowed him to arrive at his belief in a First Unmoved Mover (his term for a supreme being), or God. He mentions a "first cause" frequently, and by this he meant that there must be something that in itself is both necessary and sufficient. In other words, there must be some uncreated force prior to the universe and possessing the power to create. In addition, that being either created order or brought order into the creation. Something must have both created and started in motion what we experience around us. We know that if we place a bowling ball in the center of a room (assuming the floor is level, no wind, etc.), it will remain in that very spot until the floor rots out from under it. It requires some force outside of itself to move it. Also, for the bowling ball to get there in the first place, it requires a maker. Further, as all events require a cause, there had to have been a first cause; it follows that this cause must be something outside of our experience—something other. In short, Aristotle gave a reasoned argument for the existence of God.

What exactly does a First Cause consist of?

It is well worth our time to take a closer look at what a First Cause or First Unmoved Mover (as Aristotle calls it) is, at least as far as the attributes that must be possessed to fill that role. In order to be the creator of time and space, it follows that our being, or entity, must transcend (be outside of and not subject to) both time and space. Of course, the First Cause implies that it has had no beginning or cause of its own. It just is.

Surely one who is capable of calling these things into existence is not to be thought subsequently subject to his creation. It follows, therefore, that this entity is timeless (eternal). Further, this being must be immutable (changeless) because to be timeless (eternal) necessitates changelessness, and it must also be immaterial (spirit) because changelessness necessarily implies immateriality as all material things change. It is perhaps stating the obvious but our being must also be all-powerful because he is able to call/speak/bring forth the universe without recourse to any material cause. We'll have cause (no pun intended) to refer back to this later.

Epicurus

While it cannot be claimed that Epicurus possessed the same depth of thought and analytical ability as Aristotle, he has nonetheless managed to have an enormous impact on Western society, particularly over the past five hundred years. Born in Samos (341 BC) of Athenian parents, he went to Athens at age eighteen and studied under Xenocrates (you may remember him as Aristotle's buddy) at the Academy. (Samos is a Greek island in the eastern Aegean Sea a mile off the coast of Turkey, separated from it by the Straight of Mycale. The mathematician Pythagoras was also from Samos.) He apparently chose the Academy randomly as Aristotle was teaching at the Lyceum and Epicurus himself said that he didn't know who Plato and Aristotle were. School, or at least being taught, wasn't of much interest to the young man because he soon left to wander. Some while later, he showed up once again in Athens and set himself up as a teacher of philosophy with his own school. Epicurus boasted of his being self-taught; Cicero later commented rather sarcastically that one would have guessed as much, even if Epicurus had not stated it himself, as one might of the proprietor of an ugly house who should boast of not having employed an architect.

Epicurus

Not an especially expressive writer, he could at least speak plainly and did seem to ably articulate his thoughts. Most of his writings have been lost; but fortunately, depending on your point of view, his ideas were retained and expanded on by Lucretius (a later Roman poet, b. 99 BC) who said of him: "Glory of the Greek race who first hadst power to raise high so bright a light in the midst of darkness so profound, shedding a beam on all the interests of life, thee do I follow." This was indeed great praise, and Epicurus did have crowds of followers who loved him, and many were proud of being able to memorize his words. He seems to have been kind and amiable and the possessor of simple tastes. It is important for us to understand that when the theory of material origins appeared on the scene via Charles Darwin, it was only the blossoming of a seed planted nearly two millennia prior by Epicurus who, in turn, borrowed the basics of his philosophy from Democritas. Of special interest is Epicurus's starting point, which was the statement "Ex nihilo, nihilo fit. [From nothing, nothing comes.]" This seems like a common-sense thing to say but it may surprise you to find that Epicurus (as well as many of his modern-day brethren) quickly lost the solidity of his foundational statement.

In his view, philosophy had but one end, and that was in utilizing the intellectual tools of reason to attain a realization of happiness. Arts or sciences that did not lead to that end served no purpose. The function of physics was only to assist in ridding humanity of myths, ghosts, and legends. The highest attainment for man was the most consistent attainment of pleasure, which he believed to be synonymous with "good." Obviously, at the other end of the spectrum, is unhappiness or pain, which is "evil." The salient point of his philosophy being the exorcism of evil (pain) from our lives. Now in what did this evil consist? Epicurus believed that the overwhelming majority of human unhappiness or angst was due to their continual concern regarding their personal and collective relationship to the gods. People suffered (or were afraid that they would) when they displeased the gods and were constantly engaged in efforts to keep them appeased rather than spending their lives in peace and enjoyment. The gods held the keys to the afterlife, in which there was an entrenched belief system. In addition, how were they to know whether they should be considered worthy to enjoy bliss in the afterlife? If they were not involved in trying to keep the gods appeased, how could they hope that the beings who held their destinies would admit them to paradise when the time came?

Another cause of human pain (via anxiety and fear) was the superstitious concern over such natural events as lightning, thunder, comets, eclipses, sun, moon, stars, etc. These were viewed as either gods in their own right or efforts by the gods to communicate with humans—or both. If science were viewed in the light of freeing people from these sources of suffering, then this would bring about a certain degree of utopia on earth. Science was not, therefore, conducted with the intention of finding truth but merely as a sort of sedative—a kind of ancient Prozac. Now the quickest way of getting rid of all this unpleasantness was simply to rid oneself of the gods.

Epicurus, however, had enough philosophical training to realize that any hope in being taken seriously by his peers, would require that he not just wish the gods away but in providing a plausible, full-blown cosmology (the study of the origin and nature of the universe) to replace them. Epicurus's ideas regarding physics are of much greater interest to us in our particular journey. His benchmark is the proposition that "nothing can be produced from nothing and nothing can perish" (ex nihilo nihil fieri posse perire). He applied this to the basic building blocks of all things which he called atoms—too tiny to see but still possessive of a definite size that could not be divided into smaller units. (Epicurus built much of his "atomic" theory by using the earlier teachings of Democritus though Mr. D. was much less clear in his arguments and conclusions.) They also possess definite shape and weight but have no other properties than these. He saw space as infinite and the number of these atoms to be infinite as well. Some of the space is void as evidenced by the fact of motion, and this motion also serves to keep the atoms separated from one another.

These atoms would naturally fall straight down—though what produced the motion is obscure—forever through infinite space and maintain parallel tracks, thereby never intermingling. This was a problem because Epicurus maintained that atoms intermingling (in the right way) was the method whereby all material things came into existence. To remedy this rather severe cramp on his style, he advanced atomic *deviation*—a sort of random twitch, I suppose. Deviation would allow atoms not only to accrete into worlds but also to make possible all the other life-forms. Through this deviation, without any explanation of its cause, of the descending atoms, the law of necessity was unilaterally done away with leaving the alignment and collision of atoms to chance. (The law of necessity states that everything that has ever happened and ever will happen is necessary and cannot be otherwise. Aristotle defended this logically by stating that only one of two conflicting statements can be true; otherwise, it would permit an unallowable contradiction. He removed chance and contingency from philosophical argument.) At this point, Lucretius (several hundred years later) takes over and introduces some elements of his own.

Lucretius

In his epic poem *De rerum natura*, admittedly picking up where his hero left off, Lucretius announces that the order in nature reveals the truth of this proposition (creation through the accretion of atoms). In short, he promotes the philosophy of materialism. All things are formed by atoms. The atoms have existed eternally. They cannot perish. It is through the decay and temporary death of things that atoms are re-released back into the ether and allowed to create new things. If this were not so, infinite time would have long ago exhausted all the matter in the universe. Clearly, then, nature is immortal. There was no design (except insofar as it was contained within the atoms themselves), no purpose, no reason for the universe to be here. There are no gods, so certainly only the superstitious will continue to fear them. In *De rerum natura*, there is the skeleton of Darwinism fully eighteen hundred years before Darwin himself. Let's look at two quotations from that poem that demonstrate my point:

> For certainly it was no design of the first-beginnings that led them to place themselves each in its own order with keen intelligence, not assuredly did they make any bargain what motions each (atom) should produce; but . . . struck with blows and carried along by their own weight from infinite time until the present, (atoms) have been accustomed to move and meet in all manner of ways, and to try all combinations, whatsoever they could produce by coming together, for this reason it comes to pass that being spread abroad through a vast time, by attempting every sort of combination and motion, at length those come together which . . . become the beginnings of great things, of earth and sea and sky and the generation of living creatures.

As a result of this atomic trial and error:

> In the beginning (principio) the earth gave forth the different kinds of herbage and bright verdure about the hills and over the plains, and the flowering meadows shone with the color of green; then the various kinds of trees came a mighty struggle, as they raced at full speed to grow up into the air. . . . So then the new-born earth put forth herbage and saplings first, and in the next place created the generations of mortal creatures, arising in many kinds and in many ways by different processes. For animals cannot have fallen from the sky, nor can creatures of the land have come out of the salt ponds.
>
> It remains, therefore, that the earth deserves the name of mother which she possesses, since from the earth all things have been produced. . . . (And so) the earth, you see, first gave forth the generations of mortal creatures at that time, for there was great abundance of heat and moisture in the fields. Therefore, wherever a suitable place was found, wombs would grow.

But *how* did all this come about? What was the method or mechanism? He says:

Many were the portents also that the earth then tried to make, springing up with wondrous appearance and frame: the hermaphrodite, between man and woman yet neither, different from both; some without feet, others again bereft of hands; some found dumb also without a mouth, some blind without eyes, some bound fast with all their limbs adhering to their bodies, so that they could do nothing and go nowhere, could neither avoid mischief nor take what they might need. So with the rest of like monsters and portents that she made, it was all in vain; since nature banned their growth, and they could not attain the desired flower of age nor find food nor join by the ways of Venus. For we see that living beings need many things in conjunction, so that they may be able by procreation to forge out the chain of the generations.

And many species of animals must have perished at that time, unable by procreation to forge out the chain of posterity: for whatever you see feeding on the breath of life, either cunning or courage or at least quickness must have guarded and kept that kind from its earliest existence; many again still exist, entrusted to our protection, which remain, commended to us because of their usefulness.

But those to which nature gives no such qualities, so that they could neither live by themselves at their own will, nor give us some usefulness for which we might suffer them to feed under our protection and be safe, there certainly lay at the mercy of others for prey and profit, being all hampered by their own fateful chains, until nature brought that race to destruction.

Lucretius, De Rerum Nature, Loeb Classical Library vol. L 181, trans. W.H.D. Rouse, rev. Martin F. Smith, (Cambridge, Mass.: Harvard University Press, 1924, 1975)

So What?

The reason that I have cited these two philosophers is because they are at the core of the two world views available to us. Yes, there are only two and whichever one we choose will shade all our thoughts, decisions, and actions. One the one hand is the view that the universe was created by a supreme being—yes, a supernatural being—with intent and purpose. Our lives matter. Our decisions matter. Objective morality, good and evil, right and wrong exist as real things. The other hand holds the belief that nothing exists but matter. We are the result of a conglomeration of atoms and/or chemicals and accident. Nothing ultimately matters, and good and evil are only social constructs which the powerful (or the effective charlatan) are able to impose on the rest of us. This second option holds supreme place in the world today. It is considered the ultimate intellectual view because it is based not on superstition but science. It is this science that we are going to investigate.

I can hear some argue that they don't have a world view. From the very beginning, since the questions were first posed "Why am I here?" or "Where did I come from?" all men have had a world view; and they have all, ultimately, revolved around a natural or supernatural construction. You may have never bothered to consciously ask yourself these questions. It doesn't matter. Indeed, you may respond that "I don't care where I came from or why I'm here." This very response dictates the world view you have chosen by default, and your life will testify to it. Some people attempt to mix the two, and the result is confusion and ill-defined life parameters; and they find themselves constantly swerving back and forth over the yellow lines, never knowing which lane they belong in. It is to provide some navigational aids for such folks that I write this book.

Before we proceed, let me give you the definition of a fact: noun; a thing that is indisputably the case. "Well now, Dave, thanks for sharing that, but we already know what a fact is," I can hear you say. My point is that there are not facts that pertain to one world view and a different set of facts relating to another. Facts are facts. What we will come to see, however, is that there is an unbridgeable gap separating the way in which these facts are *interpreted*. In other words, the lens through which we view the fact will color our understanding and conclusion regarding it. For example, one person can gaze into the Grand Canyon and wonder at the millions of years it took the tiny (comparatively) Colorado River to etch it out. Another person considers the same canyon and believes that there is no way that little river ever caused such an immense hole in the ground. The Grand

Canyon is a fact; but both individuals, having viewed that fact, have reached entirely different conclusions.

Just to be clear (after all, our search is for clarity), let's take a moment and define exactly what science is and what it is not. A generally accepted definition is something like "the knowledge of or study of the natural world based on facts learned through experimentation and observation." This is known variously as operational, empirical, or hard science. This is the science that has given us our great discoveries in, among other things medicine, energy, and information processing. It is the science by which we understand how chemicals combine, what the densities of various gasses are, and the fascinating aspects of human anatomy. This science has allowed us to walk on the moon, plot the location of planets, and calculate their orbits and to (lamentably) create the atomic bomb. It is in the laboratory that we seek cures to AIDS, cancer, and Alzheimer's by trial and error, testing, quantifying, and verifying. The smart folks in the white lab jackets, rubber gloves, and goggles are engaged in this type of science, ever seeking (supposedly) to make our lives better through a new vaccine or a better mousetrap. This is decidedly *not* the type of science that we will be discussing over the next few chapters.

This other type of science is called historical science. This pertains to the investigation into origins. I don't mean the origin of the toaster oven or color TV but *the* origin. The origin of the universe and life itself are the very topics that this science attempts to grapple with. Now, by definition, this is no science at all. It is a philosophy. Origins, by their very nature, cannot be proven. For example, the earth was formed long ago. Nobody was around to test and verify how it came to be, and it is too late to do so now. *Any* world view is held as an article of faith based on the individual's interpretation of facts. It cannot be quantified and verified, and it certainly was never observed by any living person. In short, the study of origins is based entirely on presuppositions drawn from a floor plan *assumed* to exist by those who promote it. The materialistic world view was originally postulated by Epicurus and Lucretius, and it lay buried for well over a thousand years as the Christian faith spread across Europe, the Near East, parts of Africa, and beyond.

The fourteenth through the seventeenth centuries witnessed what is commonly referred as the Renaissance. This renewed interest in learning focused particularly on classical texts. All ancient writing became the craze, and with each text that was resurrected from the dust, something arose which caused enlightenment, controversy or both. Poggio Bracciolini (1380–1459) was a papal secretary under Pope Boniface IX who located many ancient manuscripts of interest in monasteries scattered throughout Germany, Switzerland, and France. One day in 1417, he stumbled upon a copy of Lucretius *De rerum natura* in a German monastery. The discovery and the poem itself were soon made known in all the academic centers of Europe. While in many ways inconvenient in a Christian culture, the ideas put forth in these documents would find themselves disseminating throughout the scholarly world.

Many great men lived and worked during this period. Among them: William Harvey, medicine; Johannes Kepler, astronomer; Antoine van Leeuwenhoek, microbiology; William Shakespeare, literature; Galileo Galilei, astronomy; Francis Bacon, the father of the scientific method; Réné Descartes, philosopher and mathematician; Baruch (Benedict) Spinoza, philosopher; Immanuel Kant, philosopher; John Locke, philosopher and political theorist; Sir Isaac Newton, Mr. Everything; Voltaire, critic; Jean-Jacques Rousseau, political philosopher; Benjamin Franklin, scientist and statesman; and Thomas Jefferson, statesman. There is no arguing the impact these men, and others, had on their day and ours.

Following hard on the heels of the Renaissance came the curiously named Age of Enlightenment or, as it is sometimes called, the Age of Reason. (I'll leave it to you to judge whether we're enlightened or not. As to reason, I can only view that proposition with emotions which vacillate between hilarity and disgust.) This age covered a time span from approximately 1685 to 1815, after which I can only assume that Enlightenment and Reason had run their respective courses and receded into history. In any event, they may just as well have named it the Age of Newton or the Age of Galileo because the impact of it can be largely traced to their lives.

As mentioned, Christian doctrine had held sway for over a thousand years, and the doctrine of the atom had been but recently resurrected. The sod in which the newly reborn atom rooted itself in these days could not have been better tilled. One major reason for this fertile soil was that many men at the time were driven to distraction by the inroads of scholasticism into the church chiefly via the teaching of Aristotle. They were agitated by the introduction of Greek philosophy into the churches institutions of higher learning and the tainting of Christian doctrine by the introduction of reason.

The view that the church took regarding Aristotle is an interesting one. The church had no problem accepting Plato because he taught that the essence of things derived from a perfect prototype that existed in heaven and was translated to earth. Aristotle, on the other hand, taught that the essence of things existed in the things themselves and could be ascertained by study and observation. In other words, Plato believe the perfect cat, tree, rose bush, etc., existed in heaven and Aristotle did not. Yet it was Aristotle that argued for God's sovereignty, and these latter day men of the church chose to despise.

So what? Well, the implication was/is that instead of looking to heavenly principles for the *meaning* of earthly things, we need to study the things themselves to derive some universal principle. Ironically, though Aristotle espoused the principle of the necessity of god, he became the poster child for modern science. And in a way (at least for a while), he was kidnapped by science because of his joy of and excellency in studying nature.

It is curious that the church decided to accept atomism as the method whereby these men hoped to fight Aristotle's influence. Little did they realize that they had willingly invited the wolf into the henhouse. Many of the men who sought to introduce the atom into science were staunch supporters of Christian doctrine and sought to use the Epicurean/Lucretian doctrine to drive old Aristotle from the precincts of the church. One of these men was Giordano Bruno (1548–1600), who had helped to extend the acceptance of the Copernican theory of celestial bodies, which was, at that time, being aggressively attacked by the church. The Roman Inquisition also accused him of denying the deity of Christ, the virgin birth, and the reality of the Trinity—more than enough to get him in serious trouble. His preaching on pantheism (the belief that all reality is identical with deity) did not help his cause. He was burned at the stake for his troubles.

Three methods of argument were at work here, and while the intent was not to endorse Epicurean ways, the effect nonetheless was just that. First, many focused on the pious, ascetic life Epicurus personally led so that they saw no harm in reading his works; after all, it was just a study of the ancient world. Second, there were those who admired Lucretius as a poet all the while pretending disgust with his philosophic leanings. However, you could not read Lucretius without coming into contact with his doctrine any more than you could touch a slug without coming into contact with its slime. Third was to appropriate materialism to Christianity. Typically, it started by saying, "Sure the universe is made up of atoms, but God created them and set them in motion." Lucretius' idea of morality could also be appropriated. Pleasure could be the highest good in this life as long as union with God was considered the highest *possible* good in the life hereafter.

Another fellow by the name of Pierre Gassendi (1592–1655) had better luck with getting the concept of atomism accepted. He managed to rub shoulders with the great scientists and philosophers of his day and found that they had a high regard for his ideas on materialism. The atoms were, of course, under the control of God (many devout men believed), nonetheless the Epicurean Jack was out of the box and would be unwilling to go back in. Gassendi's books were well received by an ever-burgeoning reading public. But the real heavy lifting associated with making materialism popular was done by Galileo and Newton.

Galileo

Galileo Galilei (1564–1642) was born in Pisa, Italy. Originally sent to the University of Pisa to study medicine, he became fascinated with geometry and was able to talk his father into letting him shift his focus to mathematics and natural history. Everything was not drudgery, however, because somewhere along the way he took a mistress, Marina Gamba, and together they produced two daughters: Virginia and Livia.

Originally a devout man of the church (he had once considered the priesthood as an occupation), he did not believe that either daughter, being born illegitimate, could make a decent marriage (he was also concerned with the cost of their dowries); so the two girls became, ironically, nuns. The couple also produced a son, Vincenzo.

Galileo was a multitalented man. He was a skilled lens grinder, mathematician, and powerful speaker. The first stir he caused was due to the observations made of the stars through his "spyglass" (the original name for the telescope). For the first time, the craters and other imperfections on the moon's surface were revealed. To discover that the moon was basically covered with dirt just like the earth was a very disturbing thing to the masses. It showed that the planets and stars were not made of some heavenly substance but of the same course stuff as our poor, fallen world. In other words, the universe was all made of the same substance: accretions of the Epicurean/Lucretian atom. Today, with this information being common knowledge, it is difficult to fathom the huge impact that this discovery had in its day. For those who had merely speculated in materialism, this came as a lightning bolt of vindication.

His spyglass also revealed a universe of more vast proportions than had previously been believed—perhaps infinitely vast (though this was carefully disguised due to the example made of poor Bruno by the inquisition). Galileo's observations seemed to support other Epicurean ideas. If the universe were, in fact, infinite in extent, then it was only reasonable to suppose that an unlimited number of worlds would exist in it. Galileo also discovered four new planets never seen before. What else might not be discovered? It made no difference that they were, in reality, moons of Jupiter. Human imagination of what may in future be discovered began to run amok. The idea of an infinite universe and infinite worlds was not something known as fact but merely a mind-based *probability*. Probability has a way, however, of becoming fact to the excitable or to those with a particular agenda.

In the Bible, Genesis tells us that God placed the lights in the firmament of heaven for light, night, day, seasons and so forth. The earth, since it was fallen and full of sin was thought to be at the center of the universe because of its being lowest on the cosmic totem—the farthest from God. As you venture out into space, you will be moved farther from the earth with its decay into the realm of beauty and eternality. Nicolaus Copernicus (1473–1543) turned this depiction of the universe upside down when he placed the sun at the center.

But Galileo's heavenly observations were not all. He viewed the world as being organized and run on a mechanical system. Ultimately, he settled on transferring the principles of geometry to the movements of heavenly bodies. He believed that the mechanical universe he desired could be graphically represented by the orbiting of planets. The planets served as geometric points, the figures they scribed in their orbits were geometric forms, and the points (planets) themselves represented atoms. As Galileo said in his *The Starry Messenger*, "Natural philosophy is written in a great Book, which holds itself at all times open before our eyes—I mean, the universe itself. But no one can understand it unless to begin with he sets himself to master the language, and recognize the characters, in which it is written. It is written in the mathematical language, and the characters are triangles, circles, and other geometrical figures." Now, of course, everyone who has taken a high school course in geometry knows that points, lines, etc., are imaginary. To say that a planet represents a point, or atom, doesn't make it so any more than to say a poached egg is roasted chicken.

What all this accomplished was to give scientists, at least in their own minds, a viable link to reality on which to hook their materialistic wagon. It cannot be emphasized enough, however, that while they could look off into the heavens, they were still unable to view the very tiny. The microscope had yet to be invented. Science could not empirically verify the existence of the atom, let alone its composition and function. The best they could do was envision it via analogy, which, apparently, was good enough not only for them but for future generations as well.

In this manner, the idea of various laws of nature became reduced to mathematical precision. It reduced nature to a closed system, which excluded any space or need for a creator (First Cause). God was not needed to make $6 + 6 = 12$ or influence the truth of the equation. Likewise, God was unnecessary to the proofs demonstrated in Euclidian geometry. As God cannot make $6 + 6 = 13$ or a square circle, neither could he change the natural expression of atomic points. Galileo faced the inquisition and was placed under house arrest for his views, but within a short time, Isaac Newton came along and was thought to be just a notch below deity (if there were such a thing).

Sir Isaac Newton (1642–1727) completed any work left undone by Galileo in cementing materialism as the mantra of science. It was not, perhaps, his mission to permanently establish it as such, but that was, nonetheless, exactly what he did. What he *was* intentional about was developing a system that explained all of nature in mathematical and atomic concepts. His two-great works, *Philosophiae Naturalis Principia Mathematica* (*Mathematical Principles of Natural History*) in 1687 and *Optics* in 1704 were rapidly inserted as the go-to references of science.

Newton

While Newton's study was based on celestial movements (initially), the ultimate purpose was to develop a universal theory for the nature of *all* things. As he outlined his famous laws of motion, you can see his intent because they were not restricted to natural bodies and motions but to include *all* bodies. It should be noted that Newton followed the methodology established by Euclid of proceeding from definition straight to law and *then* to demonstration. This method *assumes* the premises are proven and, further, that science is a process of making deductions from those "proven" premises. This further presupposed that the complexity observed in nature (at least that portion he was able to observe at the time) was reducible to . . . what? Epicurean/Lucretian/Galilean atomic points, of course.

Newton's demonstrations (illustrations) of his laws were limited to inanimate objects such as planets, billiard balls, etc.; but he applied them to animate objects as well. This co-application would only be possible if both animate and inanimate objects could be reduced to one common antecedent that was itself incapable of motion unless an external force were applied to it. In short, the atom. You see, a natural *law*, by definition, must be universal; so he simply made all bodies subject to the same principles. Obviously, there was no empirical evidence for his extension to animate bodies of his laws.

These were the three laws of motion that Newton posited:

> Law 1: If a body is at rest, it will remain at rest, or if it is in motion, it moves with uniform velocity until it is acted upon by another force.
>
> Law 2: Acceleration is dependent on the forces acting upon an object and the mass of the object. If the force is increased, the acceleration is increased. If the object has more mass, the acceleration decreases.
>
> Law 3: For every action, there is an equal and opposite reaction, or every action always reacts in the opposite direction.

Once again, the universality of these propositions would only be possible *if* all nature could be reduced to a universally small inert body. Because of this, all apparently complex bodies, all apparently complex motions, could be easily explained by their reduction to this lowest common denominator. Newton, in his *Principia*, says, "I wish we could derive the rest of the phenomena of Nature by the same kind of reasoning from mechanical principles, for I am induced by many reasons to suspect that they may all depend upon certain forces by which the particles of bodies, by some caused hitherto unknown, are either mutually impelled towards one another, and cohere in regular figures, ore are repelled and recede from one another." Atoms ruled.

As the centuries rolled on, the watchword of society increasingly became *science*. The discoveries made in engineering, medicine, and communication were certainly mind boggling. It seemed that almost before the public digested the latest invention or theory, another was plowing up the ground behind it. Scientists and inventors became our gods. They convinced us that given enough time, they could conquer any summit, answer any question. Science would explain it all. Building on those who had come before him, Charles Robert Darwin arrived on the scene and developed his theory of descent with modification via natural selection or, more simply, evolution. As we have seen, he was not the first to propose materialism, but he and those after him put flesh on the skeleton created by Lucretius.

The world of the atom, as seen by Epicurus and Lucretius, went from being something outside of Christianity to being the only acceptable view of the natural world by the seventeen and eighteen centuries. As we said, you may mark it by the lives of Galileo Galilei and Isaac Newton. There were, of course, many lives that contributed to the victory of materialism in science and some of them Christian. Perhaps foremost among them were the previously mentioned Giordano Bruno, Pierre Gassendi, and Robert Boyle (1627–1691). They were ardent haters of Aristotelianism, and they sought to use the Epicurean view as a cure for the scholasticism that they believed was hurting Christianity.

While the belief in multiple worlds continued to grow, we should not be surprised that the materialist universe of Epicurus grew as well. Indeed, Johannes Kepler (1571–1630) and René Descartes (1596–1650) hinted that an unlimited universe with unlimited worlds most probably meant that there was intelligent life elsewhere. It was inevitable that once this plurality of world's scenario (especially when espoused by many Christians) became common, that the credibility of God giving his Son for the salvation of man on one planet became less and less credible.

It should be noted, once again, that the universal application of these principles was not empirically proven; rather, it was the desired outcome that dictated its acceptance. In the decades that followed, a place for God could no longer be found in this material universe. In 1781, William Herschel discovered the planet Uranus. No matter how hard astronomers tried, however, they could not mathematically explain the erratic course of its orbit. Two men, one English and the other French, predicted, using mathematics alone, that another planet would be found at a particular place and that planets gravitational pull would account for the unusual nature of the orbit. Sure enough, looking to that very spot in the night sky, Johann Galle discovered the planet Neptune on September 23, 1846. What greater proof could science ask for the mathematical/geometrical nature of the universe than this?

This way of thinking became so ingrained that even when atoms were discovered to be finite and therefore nature could not be its own cause, the habit of continuing as if they were infinite remained. If atoms are not eternal, then the universe had a beginning. There must be an eternal cause other than nature itself that brought nature into existence. What was it? It was not deemed important to ask these questions because with the advent of Charles Lyell's publishing of *Principles of Geology* in 1830, he introduced something by which all things are made possible: time, vast, unimaginable amounts of time. Only a few years after this, Charles Darwin gifted science with a theory that made sense of it all. In 1859, he published his *On the Origin of Species*, and materialism was here to stay.

Religion was not the only area of human activity affected by these new ideas. By the middle of the seventeenth century, Francis Bacon was already developing the scientific method, which would come to control biological and natural science for the centuries that followed.

René Descartes (1596–1650)—a French philosopher, mathematician, and scientist—was educated by Jesuits. He chose to become a soldier, which gave him extraordinary opportunities to observe human nature. Descartes's philosophy was predicted on our approach to "knowing" anything must begin from a foundation of doubt. He sought to bring certainty to philosophy so looked to introduce the methods of mathematics and geometry into his system. Developing his method, he used doubt as the foundation for demonstrating his conclusions. From this premise, he developed the famous quotation, "I think, therefore, I am." Since we are able to think, our thoughts cannot doubt themselves, giving us a firm foothold from whence to explore reality. While Descartes was also a mathematician and discoursed on physics, his philosophy had practical application to everyday life. In expressing his view of the world, he exalted human reason and raised doubt to a virtue rather than its being viewed as a fatal sin. He did, however, offer two proofs for the existence of God. Again, beginning from a negative proposition, he realized that man doubted and that doubt indicated a facet of imperfection; but if that imperfection were rightly understood, the doubter would garner some knowledge of the perfect. Knowledge of the perfect, however, cannot be obtained from an imperfect mind so that the knowledge that I have of it must come from the perfect mind (God).

Another rationalist perspective during this time was introduced by a Dutchman Baruch Spinoza (1632–1677). Spinoza primarily used geometric method to arrive at his conclusions regarding the existence of God. He begins with the notion of the absolutely perfect idea of an absolutely perfect being. He maintained that to entertain the thought of anything less than an absolutely perfect being led to imperfection itself. Spinoza ended up by being a sponsor of the ontological argument which basically says that if the absolutely perfect being could be conceived, he must exist because it is more perfect to exist than to not exist, thereby rendering it *necessary* that he exists. Spinoza's conclusion, however, leads to a pantheistic idea of Godhead. Since it is not possible to have many perfect infinite beings, all things must flow by necessity through the unity of the one necessary being. The world must be viewed in its totality, and its wholeness is God.

One of the truly great minds of the period belonged to a German Gottfried Leibniz (1646–1716). A complex man, well versed in many areas of science, mechanics, and philosophy, Leibniz developed ideas that continue to shape rational thought today. While Descartes started with the premise of an "undeniable" idea, Spinoza a "perfect" idea, Leibniz came to his philosophy by building his thoughts of God around a "sufficient" idea. His first principle maintained that every event must have a cause. This is the beginning of all true propositions and can be analytically proven to be true. For example, the law of non-contradiction holds that something cannot be itself and something else at the same time and in the same sense. This "identity" is a sufficient reason for all truth claims. While Leibniz went into his ultimate truth claims in great depth, the bottom line was that it is reasonable that God exists.

While Francis Bacon and others were advancing the scientific method—Descartes and his peers looked to establish philosophically the existence of God—a man appeared named David Hume (1711–1776). Hume would change the relationship between science, theology, and philosophy forever. Hume jettisoned the idea of a supreme "anything" altogether. He based his reality on whether a statement, which purported to be fact, contained any quantities or numbers. If not, it was a nonsensical statement. Did the statement contain any experimental reasoning or idea that could be quantitatively verified? If not, it was unworthy of continued discussion. All of man's knowledge is simply the result of experience and our reflection on those experiences. Further, these experiences we are ruminating on are all arbitrary and disconnected with no direct knowledge of oneself or the world around us resulting from said ruminations. All our experience implies a contrary experience. In other words, it could have happened otherwise. Hume even held the idea of causality to be only something we adhered to based on custom. While many consider Hume to have been an agnostic, on reading Hume, one can only come to the conclusion that he was an atheist bent on the acceptance of only those things that could be quantifiably verified.

During this time of brilliant minds yet another appeared in the body of a second Frenchman: Blaise Pascal (1623–1662). Pascal resisted the skepticism that was beginning to appear in any area that had to do with God. While he believed in the value of reason certainly, the entire philosophy generated by reason at its core was, religiously and philosophically, fundamentally flawed. In his work *Pensées* (This work was completed after his death. I'm not sure how he managed to do that.), he established that reason leads to the basic truths of both science and religion. Faith is just as important as reason, and faith can only come from God. Pascal held to the conviction that materialist/mechanistic philosophy and scientific knowledge could not lead one to a fulfilled life. If "scientists" bothered to study the facts, in his view, man's good will lead to animosity and his reason only leads to irrationality. Human life is not progressing toward utopia but is interwoven with contradiction and inner confusion. Pascal explained this by describing the dual nature of man. In the first place, man is a fallen creature; history proves this. In spite of this, we also see the spirit of grace operating throughout the world and through fallen people. His argument for this scenario of human existence is also twofold. Historically, Christianity established itself and continued to grow in a manner that certainly supports its divine origin. The second premise shows that, when we experience moments of inspiration, it points to a reality that it is absurd to question.

Darwin

And then there was Charles. Darwin was born in Shrewsbury, Shropshire, England, on February 12, 1809. (On the same day, in a backwoods Kentucky log cabin, Mrs. Lincoln was delivering the baby boy Abraham.) His father, Robert Darwin, was a wealthy doctor and financier who considered himself a freethinker. When Charles was eight years old, his mother, Susannah, died; and he was sent to an Anglican boarding school in Shrewsbury along with his brother, Erasmus. At the time, it was thought that he would join his father in the practice of medicine, and with that in mind, he spent the summer of 1825 with his father providing treatment to the poor in Shropshire. In the fall, he headed off to the University of Edinburgh Medical School. Charles, however, found surgery disgusting and neglected his books. He did, however, learn taxidermy (not apparently bothered by the separating of animals from their hides) from a freed black slave named Charles Waterton.

His second year at university saw his focus change. Always interested in the natural world, he joined the Plinian Society, which was a student natural-history group that enjoyed debating other societies that had the unfortunate characteristic of entertaining orthodox religious views (e.g., creation). Since it became obvious that Charles no longer saw himself entering the medical profession, his father packed him off to Christ's College, Cambridge, to pursue a bachelor of arts degree preparatory to entering the Anglican clergy. The result of this change was that he became an avid beetle collector, apparently continuing to neglect his books and not showing much interest in the clergy. While at Christ's, he also became familiar with William Paley's *Evidences of Christianity* and *Natural Theology or Evidences of the Existence and Attributes of the Deity*. (Natural theology is proof for the existence of God as evidenced in the natural world and a popular theological apologetic at the time.)

A friend and professor of botany, John Henslow, recommended Darwin as a companion to the captain and private naturalist onboard HMS *Beagle* during its proposed two-year voyage to chart the coast of South America. His father was incensed, but Henslow convinced him that the study of natural history would be useful to Charles in his clerical career. The voyage began after Christmas in 1831 and would last nearly five years rather than the projected two. During the five years he was absent on this venture, he collected specimens and kept careful notes of his observations, periodically sending them back to Cambridge. Robert FitzRoy, captain of the *Beagle*, had given Charles a copy of Charles Lyell's *Principles of Geology* whose findings made for additional interesting observations whenever the ship struck land. He observed the various layers in rock formations, was delighted with the rainforest in Brazil, found fossil bones of huge extinct mammals, and traveled to the interior, making anthropological observations of the native peoples. All in all, to a man with a curious mind and an active imagination (stoked by Lyell's *Principles*), this was a most rewarding trip indeed.

When Darwin returned home in October 1836, he did so to celebrity. Friend Henslow had already circulated some of his findings to selected naturalists. Transmutation of species was a hot topic of discussion among scientists, and many sought to use Lyell's approach to geology as the basis of such a theory. Darwin took note. There was much pressure on Darwin to complete the publishing of his notes, speak to various groups, and articulate his conclusions. Soon he was overwhelmed and needed to resort to the country to rest. He continued to work, got married, and made progress on his efforts toward publication. While recuperating (if his activities could be called such), he hit upon the idea of natural selection as the mechanism that would drive transmutation.

Darwin spent a good deal of time researching and observing breeders of domesticated animals. It was an established fact, long before his time, that farmers were able to breed hardier stock. Dog breeders bred their hounds to produce superior strength or color and so on with cats, bulls, horses, etc. Another group of enthusiasts grew the finest roses, camellias, or orchids. The genetics involved in this process were unknown at the time, but the methods were carefully noted in breeders' handbooks, and the success (or failure) was there for all to see. This artificial breeding criterion Darwin assumed to operate in the wild with the same efficiency as on the farm. Apparently, he didn't consider it noteworthy that men, presumably possessed of intelligence, were causing these various animals to be put into a position to breed and choosing the mate. Further, regardless of color, size, or weight, dogs still produced dogs and cows still produced other cows. In addition, it was fairly easy to observe that different species possessed certain characteristics that made them more apt to thrive in heat, cold, wet, or dry environments. Those that did not possess the particular traits requisite to the circumstances or environment simply died out.

The timing of his theory, the pressure to publish because of a competing theory of Alfred Russel Wallace, the "mood" of the scientific community all combined to create the perfect storm. (Perfect storm: noun. A detrimental or calamitous situation or event arising from the powerful combined effect of a unique set of circumstances. Pun unintended.)

In 1859, Darwin published his *On the Origin of Species*. Darwin concluded that all life on earth is related through a common ancestor. The diversity of life is explained by gradual modifications within populations through natural selection and where certain characteristics are more conducive to the environment than others they are passed on to their offspring. This became more commonly referred to as descent with modification or survival of the fittest (where fitness is more closely allied with the ability to reproduce in an environment rather that physical strength or athleticism). As an example of this, Darwin said in *Origin*: "I can see no difficulty in a race of bears being rendered, by natural selection, more aquatic in their structure and habits, with larger and larger mouths, till a creature was produced as monstrous as a whale." This particular passage excited so much ridicule that it was removed from future editions of the book; it nonetheless represents rather accurately what he was getting at. Modern scientists have changed the analogy but not the basic theory.

The key factor in evolutionary theory is time. Given enough time and with the accumulation of enough modifications, natural selection is capable of changing practically any species into an entirely new one. Common pictographs of today show dinosaurs becoming birds, small tadpoles turning into whales, and, of course, apes transitioning into humans. Once available to the general public (general public: the common man with common sense and possessive of ideas of their own, I think), the theory was met with howls of both derision and hilarity. Not so among many men of science. They had been waiting on such a theory for a long time. And remember, the face of science was becoming the face of our god. It needs to be emphasized that the entire theory was predicated on *external* observations. Based on externals, it was easy for an artist to represent a fish transforming itself into an amphibian. The operations of a single cell, let alone DNA or other complex biological interactions, were entirely unknown.

By the 1870s, academia had gilded the theory of evolution with a patina of fact, which made it more palatable to society at large. Remember the film *Inherit the Wind* that came out in 1960? This movie portrayed the famous Scopes Trial or Monkey Trial, which took place in 1925 (officially *the State of Tennessee v. John Thomas Scopes*). This trial accused a substitute teacher, John Scopes, of violating the Butler Act, which prohibited the teaching of evolution in any state-funded school. This trial was deliberately staged to attract attention to the backwardness of people who would oppose scientific fact being presented to eager young minds. John Scopes was not sure that he had even taught evolution but was talked into deliberately incriminating himself so that there could be a defendant.

Now, in the movie, the defending attorney, Clarence Darrow, was portrayed by the venerated actor Spencer Tracy and the dark backward prosecutor, William Jennings Bryan, by Fredric March. The defendant, Scopes, was very effectively acted by a young sheepish-looking Dick York (of *Bewitched* fame). The outcome was a spectacular, crushing victory for enlightenment, truth, and progress. In the real trial, the finding of the court was for the state of Tennessee and creation. A historical presentation of the trial is not what Hollywood intended because the producers knew that many young moviegoers knew nothing about the true facts surrounding the case. Instead, they altered the outcome to promote their world view and to cultivate young minds. What impression was left? That to modern, educated Western minds, only ignorant, barefoot folk from the backwoods could *not* believe that life as we know it was the result of "descent with modification."

Many, however, still held to several competing versions of the theory, but by the 1930s natural selection was accepted as the basic mechanism by which evolution worked and the diversity of life on earth was explained. Scientists today have tweaked the details of the original theory time and again, but evolution remains the bedrock upon which the explanation of the origin of the universe and all life is built. We should remember as we move forward that many slight deviations to natural selection are being batted about among those in the know, but we will pursue the science as generally presented in the classroom. Besides, I have neither the time nor the inclination to go down every random rabbit hole we may encounter. I have divided the next chapters into two parts: earth science and biology. Let's get started.

Chapter Three

THE ARRIVAL OF THE UNIVERSE

A Grand Explosion

There are (or, better said, were) basically two competing versions for the existence of the universe. One is referred to as the steady state model. Prior to the midtwentieth century, nearly everyone except some who had held fast to their faith and a few friends of Aristotle considered that the universe is a closed eternal system in which new matter is continuously being created. Being forever present, it, of course, had no beginning; and, therefore, messy explanations related to *how* it came into existence could be easily avoided. Messy explanations, however, cannot be avoided when something can be shown to have *begun* because it necessitates a time when it was not.

In May of 1964, two radio astronomers made an end to the steady state theory. An antenna array in New Jersey was picking up a pretty constant buzzing that seemed to come from whatever part of the sky to which the antenna was directed. The two astronomers, Robert Wilson and Arno Penzias, were puzzled (along with everyone else). After removing any source of interference they could think of, the buzzing continued. What these two gentlemen had discovered was

CMB, or cosmic microwave background, which was reported to be the leftover thermal echo of the grand explosion that brought the universe into existence. Life was breathed into the big bang theory.

Another discovery, originally reported by Vesto Slipher in 1912 and confirmed by Edwin Hubble (of telescope fame), supported the fact that the universe had a beginning. What Slipher had discovered and Hubble defined was "recessional velocity" commonly referred to as red shift. Red shift occurs when some sort of electromagnetic radiation (i.e., light) moves to the red end of the spectrum. As it does so, its wavelengths increase, indicating that it is moving and therefore demonstrating velocity. Sort of like viewing the vapor trails that follow an aircraft. In the case of stars, it means they are moving *away* from us; the universe is expanding. Thus, *science* has shown us that the universe came into being. It had a beginning. Those messy explanations would have to be made after all.

These discoveries brought to life what has ever since been known as the big bang theory. This theory is entirely predicated on the presuppositions of naturalism and uniformity. (Uniformity states the belief that the same natural laws and processes that operate in the universe today have always operated in the same manner and at the same rates and that they apply everywhere in the universe. A *huge* presupposition.). According to big bangers, the entire universe was condensed into a single point called a singularity. (I'm talking not just all the stars, planets, dust, and so forth but the very *space* as well.) Look at it as a sort of egg sitting, I don't know where, waiting to hatch. There are significant problems with even the idea of a singularity not the least of which is the problem of where *it* came from. It represents a unilateral disregard for the law of first causes; therefore, this theory immediately reveals itself as a *philosophy*, not science.

Now, *if* we grant the existence of this singularity, what caused it to all of a sudden decide to become unstable and explode? I mean, presumably it had been sitting, minding its own business, wherever it was, somehow suspended in eternity, in a completely stable state. What changed? What caused its state to become excited? I've never seen an egg decided to boil itself. Motion, according to Newton, must have some external cause. Today it is being advanced by some scientists that a *quantum fluctuation* caused the singularity. What caused the quantum fluctuation? What the heck *is* a quantum fluctuation? Now,

I admit I had no idea what a quantum fluctuation was, so I looked it up, and this is what I found:

> In quantum physics, a quantum fluctuation (or quantum vacuum fluctuation or vacuum fluctuation) is a temporary change in the amount of energy in a point in space. This allows the creation of particle-antiparticle pairs of virtual particles. The effects of these particles are measurable, for example, in the effective charge of the electron, different from its naked charge.
>
> In the modern view, energy is always conserved, but because the particle number operator does not commute with a field's Hamiltonian or energy operator, the field's lowest-energy or ground state, often called the vacuum state, is not, as one might expect from that name, a state with no particles, but rather a quantum superposition of particle number eigenstates with 0, 1, 2 . . . etc. particles.
>
> Quantum fluctuations may have been very important in the origin of the structure of the universe: according to the model of inflation the ones that existed when inflation began were amplified and form the seed of all current observed structure. Vacuum energy may also be responsible for the current accelerated expansion of the universe.

I may not be ranked among the great minds of the world, past or present, but I frankly confess to thinking myself dumber after reading this definition. How can anyone *know* that such things existed at the beginning of the universe or that, if they existed, acted in any fashion that would have an impact on it? A dead giveaway to me is the line "according to the model of inflation," which smells a whole lot like "according to our world view it would be really convenient if this were real." Another red flag to me is, "In the modern view..." Just saying.

Another explanation advanced by some is our universe was spawned from a parallel universe, but that is not only fanciful but neglects the question altogether. Where did *that* universe come from? I'll let that one alone as it gets me laughing so hard I can't write. Suffice it to say, there are no laws of physics that explain the existence of either a singularity or a parallel universe. Not to be thwarted by the sticky laws governing physics, materialists press on just as if they had adequately answered the questions.

Remember that we mentioned previously the importance of time in evolution. Time is another reason that the big bang *must* have happened. The speed of light is put into evidence as exhibit 1 for the defense. You see, a star that is 10 million light-years away and visible from Earth must be 10 million years old—that is, if the big bang actually happened. (A light-year is simply the distance light will travel in one year. Light travels at 186,000 miles per second.) Per evolutionary-based calculations, the universe is currently figured to be 13.7 billion years old. I say currently because this figure has changed a few times.

Let's take a look at the speed of light as evidence. First, it is assumed that light has *always* traveled at the same speed. Now, if I were in an exploding universe, I would not be surprised if some things were very different at that moment than they are today. However, I was not at the great event, and neither was anyone else, so we are *presupposing* that conditions were the same then as now. There is no evidence that light has always traveled at the same speed. In fact, the opposite can be observed as we will see in just a moment.

A demonstrable counterargument to this claim that light has always traveled at the same velocity is called time dilation. In space, time progresses more slowly when passing a very large object (like a planet) because of the gravitational forces put upon it. Time actually fluctuates in different parts of the universe because of the uneven gravitational "pulls." We experience this effect on Earth. Say you were silly enough to climb Mount Everest with a clock (or with anything else for that matter) and a smarter friend was at the seashore sitting in the sun and sipping a cold drink, also with a clock. If you were to synchronize your clocks and compare them, you would notice your clock running slower. The difference noted by you and your friend would, of course, be very small; but out in deep space—well, the effects can be significant indeed. Some scientists reject this time-dilation explanation by expressing the belief that the universe is infinitely large but there is no evidence of this either. Additionally, it seems to me that to say that a finite universe (it had a beginning) contains infinite space is nothing less than an unacceptable contradiction. But perhaps my logic is fuzzy.

The use of time as a proof for the big bang, however, has problems of its own. One problem is directly related to the cosmic microwave background we mentioned earlier. It tells us that the temperature throughout the universe is stable (nearly -500 degrees Fahrenheit). Certainly, during such a colossal explosion, some parts of the universe had to have been hotter than others and residual evidence of this should be discernable yet it cannot be detected. According to people that study such things for a living, there hasn't been enough time, even with 13.7 billion years, for this stable state in temperature to have come about. (Look for the age of the universe to get older again!)

Of course, evolutionists have an answer for this, and I would be sorely disappointed if they hadn't. They postulate an inflationary period right after, almost simultaneous with, the blast, which briefly accelerated the dispersal of the stuff contained in the singularity, thereby allowing it to cool at a faster rate. This, unhappily for the theory, is complete conjecture, nay, fancy. No evidence exists to support this; it was merely invented to fill a hole in big bang cosmology. Also, it occurs to me that, having counted on the law of uniformity previously, it has conveniently been laid aside in this instance. But again, maybe that's just me.

Based on billions of years, there are observations that we should see but don't. One of these is in the formations of galaxies. Just like our own Milky Way, many galaxies in the universe are spiral shaped with long arms reaching out from a central core. If the universe were billions of years old, these galaxies should long ago have settled in to a uniform shape because since the center spins more rapidly the outer arms would be pulled in.

Our Home

At first, there was nothing in the exploding universe but energy, and this energy, over time and continued expansion, was converted into matter. Eventually, the matter began to stick together (accretion) and the first atoms were formed. (I can see Epicurus beaming with pride.) As the speeding energy continued to expand, it eventually started to apply the brakes and cool down. The energy reformed into helium, hydrogen, and lithium. These gasses formed, in turn, a nucleus (nebulae), which gave birth to our planets and stars. Now, the earliest stars became nuclear bombs and produced heavier elements (through fusion), which scattered throughout the rest of the universe. This heavy stuff began to collide and accrete, thus forming the basis for additional planets and star systems. So the theory goes, in a quick thumbnail sort of way.

It should come as no surprise that there are serious problems with this explanation. The theory is that the gasses collapsed because of the influence of gravity and that the gases then began to heat up, causing the nebula to begin spinning and forming themselves into disk-like shapes. Gases, contrary critters that they are, *expand* when heated rather than contract, however, so that they would have rescattered themselves. Evolutionists seek to solve this problem by suggesting that counterexplosions would force the scattering material back in the other direction, effectively keeping it in place, thereby allowing the star to form. This simply will not do. In order for these counterforces to happen, stars would already have to be in existence and at the supernova stage. In other words, old decaying stars would have had to be in existence while the universe was being created. Even today there are few supernovae that have been identified, and after 13.7 billion years, the heavens should be thoroughly rotten with them.

Following the we'll-figure-that-out-later formula, the theory moves on. The star continues to form, held in place by the counterexplosions, and becoming more stable, forming a baby star (protostar). The baby star continues to collect dust, gases, and other particles that happen into proximity close enough for them to be captured. When it accumulates enough material to reach a certain temperature at its core—10,000,000 degrees Kelvin (for those of you who don't know Kelvin, that's 17,999,540.33 degrees Fahrenheit, which is pretty hot, I think)—the star becomes stable because it has reached the stage where nuclear fusion will occur. (Odd to think that nuclear fusion would make something stable!) Nuclear fusion is a reaction in which two or more atomic nuclei come close enough to form one or more different atomic nuclei and subatomic particles (neutrons and/or protons). The difference in mass between the products and reactants is realized in the release of large amounts of energy—*really* large amounts of energy. Stability in the star arises when the outward pressure of the gases becomes equal to, or balanced by, the gravitational pull. Once again, the "how" of all these effects taking place in this fashion is based on principles that can be observed *today*. It is assumed that the same conditions existed 13.7 billion years ago during this extraordinary (to state it mildly) period of chaos. Moving on.

It is believed that the formation of our solar system was similar to other typical systems in the universe, and this process has achieved the title of the nebular theory or nebular hypothesis, which seems a contradiction as I'm not sure how it can be a theory and hypothesis at the same time. It depends on who's talking about it, I suppose. In a nutshell, nebular theory consists of the following: Stars gather in dense clouds combining GMD (giant molecular clouds) and molecular hydrogen. These clouds are constantly attracting additional matter and create denser "clumps," which begin to spin and collapse on themselves. The star thus created is surrounded by a "protoplanetary" disk, which continues the process and, in certain circumstances, creates planets. (Such a one is our very own solar system. Our sun took about a million years to become "operational." I don't have the name of the scientist who was present when the sun was turned on.)

This protoplanetary disk is a veritable space vacuum cleaner gathering to itself any material that passes within reach, always feeding the central star. Eventually, the disk cools off, and rocks and ice may begin to accumulate to form "plantesimals" (think planet in utero). These plantesimals will collide and adhere to one another, creating planets. The planets formed from heavier materials will be closest to the star while those made from lighter materials and gases will be farther from its gravitational influence.

Nebular theory has its own unreasonable detractors. For one thing, they say, it is assumed that particles would stick together instead of just bouncing off one another and rocketing off into space, creating an unending game of pinball. If two wooden baseball bats are smashed together, they don't stick together but break apart. But remember, initially, during the formative time, we are not talking about things the size of baseball bats but very small particles of dust and such. Stars and planets, let alone systems, are said to start from gathering this type of material.

Another problem has to do with the law of conservation of angular momentum. That's a mouthful to be sure, but anyone who has watched figure skating has witnessed it in action. You know that movement at the end of the performance where the skater begins to spin with their arms stretched out to their sides? As the skater begins to pull their arms in toward their body, they begin to spin faster and faster. There you have, basically, the law of conservation of angular momentum. This is important because, if nebular theory is correct, the sun should be spinning much faster than it is. Our sun possesses nearly 99 percent of the solar systems total mass but only 2 percent of its angular momentum (a good thing too, or we wouldn't be here). Evolutionists have come up with the idea of "magnetic field braking" to compensate for this minor aberration but haven't yet decided among themselves how it's supposed to work.

Let's talk for a minute about Earth, our beautiful, beloved home. It's special, as I'm sure you'd agree, and requires special handling. Early in the life of our solar system, Earth was formed by the collisions of many bodies, which became a molten mass due to still other collisions with particles that had been orbiting the forming sun. As the earth started to cool, layers were created because of the heavier materials settling toward the bottom (the core) and the lighter ending up closer to the surface. During this cooling process, our atmosphere was formed and added to by gases being emitted from volcanic eruptions. Our original atmosphere was supposedly comprised of methane, carbon dioxide, sulfur dioxides, nitrogen, and ammonia with oxygen being conspicuous by its absence. (The absence of oxygen is necessary to evolution as will be seen later.) Its absence is explained by stating that since the planet was not formed, there were no oxygen emitting life-forms available to generate it as there are today. Another convenient setting aside of uniformity. If ever there was an example of using the theory to argue for the theory, this is one. A by-product of the volcanic emissions was the trapping of water vapor in the atmosphere. (By the way, doesn't the forming of water require oxygen?) As the earth cooled further, the water was released in the form of rain and the oceans and seas were formed.

Probability

Before we go further, let's talk math. I can hear the groans, but it is necessary to try to satisfy those of us who put great store in numbers. After all, mathematics form a large part of the calculations using in arriving at the schemes that we've been discussing. I'm sure that you've heard the term "It happened by chance" at least once in your life. What does that really mean? To say that something can *happen* because of chance infers some sort of power or motive force to chance. Does chance possess any kind of inherent power? No, chance is merely a label or term that we assign to mathematical probability. So let's take a look at what mathematical probabilities look like.

Probability is the system we use to determine the chances of occurrences coming about randomly. The most commonly used example to demonstrate probability is the coin flip. We know that the "chance" of flipping heads is one-half or 50 percent, but what are the chances of doing it twice in a row? Mathematically we take our 50 percent and cut it again. In formula form it looks like $1/2 \times 1/2 = 1/4$ or 25 percent. Probability drops pretty quickly, doesn't it? What are the odds of flipping heads ten times in a row? Without taking the space to produce the formula, it turns out to be one chance in ten million. Science, in kindness, has adopted the method of stating huge numbers in *scientific notation* to cut down on all the zeros. For example, 1 chance in 10 million looks like this: 10^7 where the seven (the exponent) represents the number of zeros to follow the one, in this case 10,000,000.

The numbers we're talking about in evolution would be unintelligible without the use of scientific notation. It does tend to remove, however, the pure effect of just *seeing* the gargantuan nature of the very numbers themselves. As an example of what I mean, does 10^{45} have the same impact on you as 1,000,000,000,000,000,000,000 ,000,000,000,000,000,000,000? I think not. This, by the way, is the number that results merely from the chance of flipping heads 150 times in a row. That's *one* chance in 10^{45}. In the case of the origin of the universe, we're not talking flipping coins. We're talking the advent of events involving a *far less* probability than that of flipping heads 150 times.

I read of a gentleman once, I don't recall his name, who stated the probability or chance of the universe coming into existence via the big bang like this: Imagine yourself viewing a scrap yard from a hilltop. There are, say, one hundred acres of scrap metal and other rubbish in this yard. Now, for no known reason, a huge explosion occurs (don't worry, you're in a safe place); and when the dust and debris settle, there before you is a fully functioning shiny ready-to-fly Boeing 777 jumbo jet that has assembled itself via the explosion and the scrap. (Some die-hards may say, "Well, it isn't impossible." I have no words for those people.) Personally, I'd rather stake my chances on flipping heads 150 times in a row. Moving on.

This is a good place to insert some interesting facts regarding the likelihood of life being sustained on our planet (or any other). This is referred to as either the weak anthropic principle or the strong anthropic principle. They are defined as follows: weak (conditions that are observed in the universe must allow the observer to exist) and strong (the universe must have properties that make inevitable the existence of intelligent life).

What is the evidence for the anthropic principle? Well, to begin with, the universe is expanding at just the right velocity. If expansion were any faster, the matter would be spreading too rapidly for any galaxies to form. That would present a serious problem for *us* as we inhabit a galaxy, I think. If the reverse were true, if expansion were slowed, all the matter would have clumped together into a gigantic ball before any individual stars could form. Another problem for us because—you guessed it—no galaxies, no us. The chance of this having turned out just right randomly is 1 in 10^{180}. (Once again, I borrow the math.) Other points of interest:

- We have one moon, which provides us with stable tidal forces. More than one moon would create tidal chaos.
- The tilt of the earth's axis controls our seasons and climates. All forms of water (liquid, vapor, ice) are necessary to the maintaining of life in its various forms.
- Earth is just the right distance from the sun to keep us warm but not fry us to a crisp.
- Our sun is situated in the proper position to promote photosynthesis.
- The sun is just the right size to distribute energy on Earth in the correct proportions.
- We have the right mix of oceans to land to keep the global temperature stable.

- Believe it or not, the distance of Jupiter from Earth and its mass and its gravity are critical factors for life on Earth.
- Our oxygen and carbon dioxide levels are balanced to promote animal and plant life.
- We have phosphorus in the earth's crust, which is critical for maintaining healthy bones and muscle. Selenium is also available in the crust, which is an important antioxidant.
- Volcanoes spread important nutrients into the soil.

There are many others that could be listed, but just from this, we can get the idea of just how unusual our planet is. One might say that it was made so with us in mind.

How Things Are Dated

Evolution has not stopped but has continued to roll on in spite of the problems involved in getting it off the ground. We will pick up our discussion with our feet finally firmly planted on good old terra firma. Before we begin, let me say that I have intentionally, thus far, avoided any meaningful reference to God in this chapter because I wanted us to consider evolution solely on the merits of the argument; indeed, we will continue to do so. However, at this stage, it is important to point out a major difference between the two world views that has been present in the background all along though not specifically mentioned. A world view that conforms to the belief that God created the world (First Cause) has no need for long periods of time. Evolution, on the other hand, relies on billions of years to make its case, the basic contention being that *anything* can happen given enough time (like our Boeing 777, for example).

Our Earth, we are told, was originally a molten ball, which cooled and retained water in the atmosphere, which was finally dumped in the form of rain and created the oceans. It is calculated to be 4.5 billion years old. During much of the time, mountains have risen and fallen, and rocks have been recycled from the surface to the core and back again. Ice ages have come and gone, and local climatic catastrophes have occurred. Scientists have developed methods to date the absolute age of everything. Again, the principle of uniformity is conveniently brought back into play. Just a reminder:

this principle involves the idea that processes that occur today have always operated in like fashion and are universal. The problem is that uniformitarianism cannot be scientifically verified as there is no way to definitively measure the rate at which events happened in the past.

Naturally, since man has walked the planet, observations surround any seeing and thinking person who chooses to make note of them. From these observations, man has arrived at certain facts. Once again, understand that there are no such things as "materialist" facts and "supernatural" facts. A fact is a fact whether ascertained by an astronaut or an aborigine. It is in how we *interpret* these facts that the difference comes in. *This* is what shapes our individual world view.

The earth, you will remember, began as a molten mass and generated our atmosphere as it cooled. Volcanic gases added to the atmosphere, creating water vapor, which finally fell to Earth in the form of rain, thereby creating the oceans and seas. The data accumulated over the years has presupposed the *fact* of evolution. Initially, fossils found in the layers of rock (which in many places are easily discerned) were used to identify similar fossils in rock layers in different locations. This resulted in the original *geologic column*. Charles Lyell's *Principles of Geology* began to postulate the idea that vast periods of time were necessary for these rock layers and fossils to form. You may recall that Darwin had studied this book, and he adopted Lyell's ideas in developing his theory of descent with modification in the field of evolutionary biology.

It was assumed that less-developed life-forms would be found at greater depths while the more advanced would be closer to the surface. The actual ages of these materials were not associated with the geologic column until radiometric dating was developed in the 1930s to 1940s. General expectations anticipated finding similar fossils at similar depths in disparate places like Colorado and China. This is where *index fossils* come in.

An index fossil is one that has been identified as *belonging* to a particular rock strata. It works (or not) something like this: *if* you know that the fossil is found in a specific stratum and *if* you know that strata formed three billion years ago, you know the age of the layer wherever you find that fossil. Utilizing index fossils can present its own set of problems (for example, all the *ifs* in the previous statement); and to be fair, even evolutionists are not in agreement regarding their suitability for use in the dating process.

Another problem comes when it is not agreed whether differences in a specific fossil are adaptations within a particular species or a distinct species. Everyone agrees that life-forms can adapt to an environment for many reasons, but a troglodyte is still a troglodyte and a dog is still a dog. In addition, fossils are *predetermined* to have been produced by evolution because, of course, evolution is fact. Another example of using the theory to prove the theory. The is referred to as a circular argument or 'begging the question.

Catastrophic events are part of the evolutionists' old earth process though considered to have been local events only. The notion of a global flood, for instance, is not considered even remotely possible. However, we see in the rock layers and fossils that the idea of a global flood is a much better fit with our observations than the argument for vast amounts of time.

The Grand Canyon is an excellent example of this and of how facts can be interpreted differently depending on your view of the world. Could the relatively insignificant Colorado River really have carved out the Grand Canyon over millions of years? Is there an alternative that is not only viable but actually a much better fit for the available evidence? The Grand Canyon is from 4 to 18 miles wide and about 220 miles long. Many geologists admit that they don't know exactly how the canyon was formed, but the accepted, default argument is that of erosion caused by the river.

First, let's consider the "facts" of the evolutionary explanation. Around the canyon, there are massive limestone deposits, which show up in various places. Shouldn't this limestone also contain common river sediments if it had been put there by a river? They don't. These limestone deposits must have been there *before* the river started to flow. In addition, there are water-transported rocks on *both sides* of the canyon. The flow of the river could not have placed these rocks in such a position as the canyon itself would have prevented it.

Common sense would also cause us to question the ability of the Colorado River to carve out a land feature so massive that it can be easily spotted from space. We have to ask, why haven't much larger and faster-flowing rivers done the same? Why isn't there a canyon down the center of the United States caused by the mighty Mississippi

River? Why isn't there a similar carving out in Egypt along the Nile or in South America following the course of the Amazon? Further, where did the hundreds of side canyons that enter the main canyon (perpendicular to the river's flow) come from? And, by the way, where did all the dirt go? Eight hundred cubic *miles* of dirt has been removed to form the canyon. If it was eroded by the Colorado, why isn't there a huge delta where the Colorado empties into the Gulf of California?

Another unexplained feature observed in the Grand Canyon (and around the globe) is the great unconformity. Uniformity, in geologic terms, is the contact among sedimentary rocks that are significantly different in age or among sedimentary rocks and older eroded igneous or metamorphic rocks. Unconformities represent gaps in the geologic record—periods of time that are not represented by *any* rocks. In other words, there is stuff that's not there that should be and there is no evolutionary explanation for it. The Grand Canyon offers some excellent examples that can be easily seen that describe what we're talking about. The time period that is "missing" in the rock layers is between 200 million and 1,200 million years. That seems significant to me, but I don't know. I'm only a writer. An unconformity also divides rock layers possessing familiar fossils from those with no fossils or only fossilized bacteria. (Remember what we said about the inaccuracies involved in using indexing fossils?)

What would the process of a global flood look like in creating the Grand Canyon? Let's see. As the floodwaters drained from the thicker (through sedimentary depositing) and elevated (floating) continents, lakes would be left behind in basins. The Rocky Mountains would settle into the earth's mantle, causing the Colorado Plateau to be hydraulically lifted an average of 6,200 feet. Sitting on top of the plateau were two enormous lakes: Grand and Hopi. At some point, Grand Lake breached its boundaries, causing Hopi Lake to breach as well. When this occurred, the waters of the combined lakes plunged off the western edge of the plateau and swept away the Mesozoic rock layers (the great unconformity) and carved out the Grand Canyon in a matter of weeks. Hundreds of smaller lakes lying about the edges of the forming canyon also breached and joined the torrent rushing southwestward, thereby forming the side, perpendicular canyons joining the main chasm.

There are also some additional formations that lend weight to the global flood theory. The Tapeats Sandstone has places where the layers are bent to as much as ninety degrees. It makes more sense to believe that the deposits were soft when the folding occurred for if the layers had been solid at the time of folding there would have been stress fractures. There is no evidence of fracturing. There are numerous other "bent" deposits around the world.

In 1980, Mount Saint Helens provided us with a real-time example of the awesome scouring effect caused by large volumes of rushing water. When the volcano erupted, the debris and volcanic ash that were emitted caused the Toutle River to block up, forming a natural dam. A few years later, the lake that had formed breached its banks and scoured out a canyon 140 feet deep in no time at all. The side canyons also look much like those in the Grand Canyon. (This "new" canyon is called the Little Grand Canyon of the Toutle River.) Nearby, Loowit Canyon was also cut to a depth of 100 feet and that out of solid rock. This sort of rapid water scouring is observable in many areas around the globe.

We've mentioned fossils several times, but what exactly is a fossil? Fossils are the preserved remains of a once-*living* organism. Science leaves us feeling as though they are somehow able to date the creation of rocks. Not so. The dates they arrive at for rock layers are based on the fossils they find in the layers. Date the fossil, date the rock. Those are the index fossils we talked about earlier. The theory assumes low-level life-forms are deep and higher-level life-forms are closer to the surface. This, somehow, is used to prove the theory of descent with modification. However, many fossils turn up in places where they shouldn't be. (Bad fossils!) What happens when a fossil goes astray? They are merely set aside as aberrations. They don't count.

The fossil record, or lack thereof, is one of the main sticking points in the Darwinian theory of biological evolution. Darwin himself pointed out the unfortunate lack of fossil evidence for the theory but placed his hope in future scientists finding what he could not. Guess what? One hundred fifty years later they have found zip.

Let's consider this from a common-sense perspective. Just taking mammals into consideration, there are currently some fifteen thousand species identified (assuming that they have been identified correctly) on the planet. Understand that descent with modification requires the changing of one life-form into another life-form by a long series of miniscule changes. I believe that a conservative estimate would require literally millions of ancestors to today's mammals. We shouldn't be able to toss a dog without it landing on some transitional fossilized predecessor of the camel or sheep or mongoose. There should be fossilized three-legged catfish lying around everywhere. Yet for all the digging, grubbing, and sweating, not one fossil that can be called a transition has been found. And believe me, they have been looking in earnest for 150 years.

What, in fact, do we find? Worldwide, we find massive layers of sedimentary rock containing billions of the fossilized remains of plants and animals all fully formed. One example that you can visit is the Dinosaur National Monument in Utah a.k.a. The Wall. The Wall is made of sandstone (sandstone is a common sedimentary rock formed from sand-sized grains of quartz or feldspar cemented together with either silicate minerals or minerals such as calcite) and exposes a graveyard of snails, logs, freshwater clams, diplodocuses (one of the long-neck sauropods), camarasaurus (another long-necked sauropod that looks suspiciously like the diplodocus), and stegosaurs (one of the armored dinosaurs), and many other organisms—all buried *together*. How did this happen? The National Park Service says that the explanation lies in a waterhole drying up about 150 million years ago so that all the poor critters died of thirst. After the drought ended, the river filled up again, covering the deceased with sediment and the fossilization process started.

As the last resort of simple folk, let us look once again to common sense. I live in Southwestern Virginia in the Appalachian Mountains where men are men and roadkill is as common as apple butter. Have you ever observed what happens to roadkill (assuming it does not end up in somebody's cook pot)? If it is summertime, the animal, if not squashed outright, will blow up like a furry balloon and burst. Bacteria will begin the process of decomposition almost immediately. And let's not forget the scavengers! I believe the state bird of Virginia (at least my part of it) should be the turkey vulture and not the cardinal. While not as pretty, they are certainly more numerous when something has died. (In wintertime, the process is somewhat slower.)

For there to be anything left to fossilize, the animal, organism, etc., must be covered quickly or the aforementioned bacteria, scavengers, etc., will consume it. At The Wall, many of the fossils are of freshwater clams. Interestingly, some are open (dead) and some are closed (alive). If there had been a drought so severe that all those animals died together, *none* of the clams should have been buried alive. (In Wyoming, The Green River Formation contains an even more unusual grouping of bats, fish, birds, and insects all fossilized together.)

There are other pieces of evidence that are advanced by the evolutionists. One of these, briefly, is the coal and oil deposits that are scattered around the earth. The standard claim is that these deposits were formed by the decomposition of plant and animal life over millions of years. They began as swamps and/or shallow seas that were gradually overlaid by the ensuing deposits. However, the fossil record does not fit this scenario. Besides, coal and oil can be created in the laboratory in a matter of hours.

I mentioned radiometric dating. What is that? It is the use of ratios of isotopes (Isotope: any of two or more species of atoms of a chemical element with the same atomic number and nearly identical in chemical behavior but with differing atomic mass or mass number and different physical properties. Clear it up any for you?) that are produced under radioactive decay to calculate the age of the object being tested. It is based on *assumed* rates of decay and other assumptions.

One common means of radiometric dating is the carbon-14 process. This method can be used on any object that was once living and contained carbon. I'll put it in context by using a good example that I read: telling time by use of an hourglass. An hourglass is an excellent time device if you can ascertain particular components with *certainty*: (1) you need to *know* the amount of sand contained in the top portion of the glass, (2) you need to *know* the rate at which the sand travels through the funnel, and (3) you need to *know* that the amount of sand in either half of the glass has not been altered. If any of these items is not known, you have a useless hourglass (useless as anything other than a paperweight or nutcracker).

Radiometric dating is predicated on similar assumptions. When radioactive isotopes decay, they form isotopes of two different elements (refer to above definition). The beginning decaying element is referred to as the *parent* and the ensuing element is called the *daughter* (don't ask me why). The time it requires one-half of the parent to decay into the daughter is called the half-life. As with our hourglass, if certain things are *known*, you can calculate the time the parent started to decay.

Example: On day one, if you had 1 gram of carbon-14, in 5,730 years of decay, you would have 1/2 gram of Carbon-14 remaining; after 11,460 years 1/4 gram, etc., to zero. Once again, our results depend on accepting some very important assumptions.

1. The rate of decay is known and has remained constant.
2. There hasn't been any change in the amount (+-) in the parent or daughter.
3. We know how many parents and daughters were present when the rock was formed.

Problems occur when we think that we *know* any of these things. First, the method assumes the known decay rates of today have been the rates of decay over the past millions (or billions) of years. Admittedly, the rates have been stable for as long as we've been measuring them, but that has covered a basically irrelevant period of time in the evolutionary scale. There are, in fact, several good reasons for not believing carbon-14 to be an accurate method for determining ages. For instance, knowing the rate of decay of carbon-14 today, coal (supposedly laid down hundreds of millions of years ago) would have "bled out" all its carbon-14 ages and ages ago—it hasn't. The same is true of diamonds (supposed to be billions of years old). Diamonds should not contain any carbon-14 either, but they do.

Another consideration surrounds change in the number of isotopes within the rock when it was formed. No account is taken of the effect of ground water or other types of weathering that have been passing over and through the rock for millions of years. Like other aspects of evolution, it is, in reality, impossible to know the original content.

There are other methods of radiometric dating that have been developed, but they are all founded on the same or similar assumptions. We have no reason to believe that any of them produce the "absolute ages" that some scientists claim.

The next major bone of contention is the global flood of Noah. Mention the name of Noah and watch materialists bloat up. All the minor issues I will leave for another time, and we will focus on whether or not a global flood could have occurred. Quite a number of books have been written that testify to a flood of catastrophic proportions but on a regional scale. As if you could be confused about what the Bible said in Genesis 6–9 or that, like the proverbial fish story, it just grew to such proportions over time by the constant embellishment of ancient flood myths. A worldwide catastrophic flood is remembered in Sumeria; Mesopotamia; Greece; Ireland; Wales; Finland; the peoples of the Kwaya, Mbuti, Maasai, Mandin, and Yoruba of Africa; China; India; Korea; Philippines; Polynesia; Hawaii; and the Hopi, Comox, Inuit, Canari, Inca, Mapuche, Muisca, and Tupi of North and South America. This is only a partial list. To be sure, the details of each are different but the overarching idea is the same as the biblical account. In addition, the biblical account is by far the best articulated and most complete of them all. It would stand to reason, if reason be accorded its proper place, that the flood legends were remembered *from* the biblical account and did not *precede* it.

How could such a global flood have happened? The Bible says: "The same day were all the fountains of the great deep broken up, and the windows of heave were opened. Gen. 7:11b

Three distinct mechanisms combined to cause the flood. We witness examples of the "fountains of the deep being broken up" today. Whenever a volcano erupts, a mudslide occurs or an earthquake happens under water; tidal waves and tsunamis follow, sometimes with horrifying results. These, however, are isolated incidents. Imagine the entire ocean floor breaking apart simultaneously! To add to the chaos, vast amounts of molten rock would start to escape through the fractures, meeting the colder ocean water, superheating it. This superheated water would rapidly evaporate, collect in the atmosphere, and produce torrential rainfall, intensifying as more and more lava escaped. We are told that the heaviest rain fell for forty days and forty nights, but the onslaught of the catastrophe would have continued and deposited eroding soils and rock debris (along with a lot of nastier stuff) into sediment layers worldwide just as we see. The Bible tells us that the waters "returned from off the earth," which indicates an uplifting of the continents, the ocean bed sinking, and the sea floor spreading out. All this movement would have spread a huge amount of sediment and other debris all over the globe.

There are many pieces of evidence visible today that display the marks of this worldwide geologic catastrophe. Two, located in Wyoming, are the Heart Mountain and South Fork Faults. As we covered previously, the mantra has long been that millions of years are required for the creation of various types of rock formations. There is, however, increasing evidence that these formations could have taken place in a matter of weeks, even days. (Actually, the evidence has always been there, but we have begun to look at it differently.) Heart Mountain and South Fork are considered the bedrock (pardon the pun) of "millions of years" geology. The enigma surrounding their formation has been debated for over one hundred years. First, we will consider Heart Mountain. (By the way, a geologic fault is a planar fracture of discontinuity in a volume of rock, across which there has been a significant displacement as a result of rock mass movement.)

Most faults are rock movements of comparatively short distances—short being some dozens of feet or less. The Heart Mountain Fault (considered a *superfault* because of its mass) is more than forty square miles in area and one thousand feet thick or more. This incredibly massive hunk of rock moved fifty miles at a rate of thirty miles an hour down a slope of only two degrees! How this could have happened continues to puzzle most geologists. One reason for the mystery is that the older layers of rock are on top of the younger ones, which, of course, is not what is expected. Investigation into this anomaly has led scientists to conclude that the large mass of rock moved rapidly, causing pore pressures in the rock to build up, providing a buoyancy to the mass. Observed in a thin layer under the mass of Heart Mountain Fault is a layer of glass-like substance that, to be there, had to be heated to over one thousand degrees by friction, I presume. This could have served as a kind of lubricant. There are no other explanations for the Heart Mountain Fault being there that I have heard that make sense.

The South Fork Fault should really be considered as having occurred concurrently (or very near concurrently) with the Heart Mountain Fault. Through the investigation of hundreds of core samples, most of which have been taken by petroleum companies, the evidence supports a fracture because of water pressure. During the late flood (Noah's, that is) or shortly thereafter, volcanic activity caused the newly exposed rocks of the Heart Mountain Fault to completely sever and slide off toward the southeast, ramming into Rattlesnake Mountain and causing part of Heart Mountain to shear off toward the south. At about the same time, South Fork, having gone through the same processes and assisted by kinetic energy supplied by the Heart Mountain movement, broke away to the southeast, trailing behind it several of the large blocks associated with Heart Mountain.

This is, of course, a simplified version of all the activity that went into creating these two superfaults. There are several full geologic reports that can be read and followed pretty closely by the uninitiated, though you will probably need to look up a few definitions of terms (as I did).

In addition to this type of activity, there was another very large contributing factor. I'm sure that when looking at a map of the world, you have noticed that the continents seem to be pieces of a puzzle that could easily be fit together. Geologists all over the globe now believe that all the landmass on Earth was once joined into one enormous continent that they have named Pangaea. During all the uproar of the flood, as mentioned above, when the fountains of the deep were broken up, Pangaea was ripped apart and, with the rising water, went scooting across the seas. In addition, when the supercontinent broke into pieces, unbelievable volumes of lava also blew out from inside the earth. Hot rocks expand and new volcanic rocks on the ocean floor rose, thereby further raising the level of the sea and pushing the waters across the now-separated continents probably at incredible speed. (There are visible on the eastern coast of Canada and the British Isles fossil-laden sediment deposits and volcanic rocks that match, indicating that they were once joined.) As an aside, it should be noted that the continents are still moving at the rate of about one to five inches per year but at the initial breakup they were moving much faster. Imagine holding a buoyant object underwater and then releasing it. The object will blast out of the water and will not stop while it has energy to move. Multiply the example previously given by vast mass and an enormous amount of water, and you can envision how quickly the separated continents moved away from one another.

There is additional corroborative evidence that this is indeed what happened. The earth's crust has become focused into an incredibly detailed picture because of the energetic (to say the least) global search for oil. The pattern that developed had mystified geologists until the 1960s when it began to clarify. In 1963, a paper was introduced that proposed that the sedimentary layers of rock bearing fossils had been deposited across the North American continent in four "packages" called *megasequences*. Megasequences refer to a continent-wide depositing in the following order: The base is a flat surface scoured of debris called *unconformities*. On top of the unconformity is a layer of boulders and large rocks (conglomerates) on top of which is a layer of sand, followed by mud and on top limestone.

You will notice that the size of the objects decreases toward the top of the layer. This indicates that the energy produced by the water also decreased. From the cores drilled for oil exploration, four such megasequences have been discovered. Geologists have been discomfited because there is no practical explanation for them using "millions of years" geology. The rapidly rising and falling waters of a global flood, however, account for them very nicely.

Many books have already been written about the problems of "spontaneous" creation beginning billions of years ago. Evolutionists also continue to try to explain the problems inherent in their theory by using a process of constant massaging (maybe I should say resuscitation). It is not the purpose of this current work to exhaustively rehash all the pros and cons as presented by the respective advocates of the two world views. I have included a list of source material and suggested further reading for those who are interested at the end of the book. Instead, I have only discussed some *fundamental* problems with evolution that serve to render the big bang portion of evolution unacceptable. At least to me, if the foundation is rotten, I don't want to invest in the house. In spite of the difficulties, however, materialists have no problem continuing their saga with the origin of life, which, when compared with the problems in big bang cosmology, is like comparing a tricycle to the *Starship Enterprise*.

Chapter Four

IT'S ALIVE!

We are all familiar with the scenario: a pool of water (perhaps bubbling for effect) in a hazy, tropical-like location, which contains all the right stuff just waiting to come alive, when, all of a sudden, lightning strikes and *WHAM!* life begins. The idea, chemically speaking, is that this pool contained all the necessary amino acids, which, when electrified (or something), formed into proteins, which then formed a living cell. A supposed *simple* single-celled organism. Makes life seem like not such a big thing, right? Well, I can assure you that getting lifeless materials to bind together and the new things in their turn join and *that* come alive is a very big thing indeed.

These amino acids are fascinating little guys. For one thing, they *can't* have bound together in some primordial swamp for the simple reason that water acts as a sort of solvent that prohibits them from binding together. (We can thank real science for this information.) This is exactly opposite of the reaction that evolution calls for. As if that little problem were not significant enough, there is a much larger one to be dealt with. To begin with, there are only certain amino acids that can be used in life generation, and these come in left-handed and right-handed models! This is called chirality, which means handedness. Molecules can *look* like they possess the same properties and elements but still have a different structure. What difference does this make? (Try tying your shoes with your hands swapped around!) These individual right- and left-handed molecules are mirror images and are referred to as optical isomers.

The problem is that proteins and DNA possess a 3D shape without which life simply wouldn't work. Now the exact amino acids needed for life, consisting of the exact number of left- and right-handed optical isomers, possessing a 3D shape would all have to randomly appear and *then* align themselves *perfectly* to form a single strand of DNA. If a method existed to create left-handed amino acids, how would it be *known* that one to create right-handed ones was also needed? Can amino acids think and plan? How, with no sample to copy, would *it* know how many of each to create? This would require some sort of mechanism to create the left- and right-handed optical isomers and to control their number and alignment. *Controlling* mechanisms do not randomly appear in nature. The single cell that popped into existence would require all this to be previously known and coded into their DNA. (One single-cell organism, *Oxytricha trifallax*, contains sixteen thousand chromosomes—DNA strands! It seems to me that the first cell would have had to be a monster containing billions of chromosomes because all the mutations necessary to come up with the 1.5 million species known today would have required them because of the loss of information resulting from the mutations. But more on that later.)

Giving Darwin something of a pass, it is only fair to say that he did not specifically advocate the idea of this chemical evolution though it was a very short hop to it. A champion of Darwin's, one Thomas Huxley (1825–1895) had no problem proclaiming to the world the reality of life popping forth from nonliving materials. Known as Darwin's bulldog, he had more to do with the acceptance of evolution among scientists and the general public than any other man of his day. Interestingly, Huxley once argued against early evolutionary theories as presented by Robert Chambers and Lamarck on the basis of insufficient evidence (!).

As we leave behind the difficulty of nonliving amino acids springing into life (as evolutionists do), let us look at how the single cell we now have becomes something else. Allow me to digress for a moment, please. Assuming our brand-new single cell is happily floating around in its pond (possessing a method to reproduce more of its kind), what made it *decide* to become something else? I mean, he has the pond all to himself, there aren't any predators, so what's he so anxious about? Is bigger, brighter, sleeker, and faster built into the chromosomes? To want to become something else reeks to me of desire—the *desire* to be bigger, brighter, sleeker and faster. To desire is to think of something to be desirous of. Can our new cell do that? Just thinking out loud.

The earliest-known fossils are of organisms that are oxygen-producing, multicellular bacteria. It follows that from the fossil record, we have no evidence that there was ever a time in Earth's history when oxygen was not present. You may recall that evolution's first atmosphere was noticeably void of oxygen. In other words, there is no evidence for the existence of our little single cell. Nonetheless, as these bacteria continued to evolve, they began to develop more complex structures internally (organelles) and retained them within a membrane. The cell formed a nucleus, also enclosed in the membrane. This nucleus was enclosed along with selective proteins that would be able to maintain the integrity of its DNA.

One theory that is advanced for the appearance of these organelles is much more interesting and dramatic (especially for those with a thing for the macabre). It is suggested that the material inside the cells are really endosymbionts. Endosymbionts are organisms that were "eaten" (or somehow consumed) by the host cell but never digested. These ingested cells were so happily accommodated in their new abode that they reproduced in harmony with its host, thereby benefiting them both. Naturally, DNA passed from the endosymbionts to the nucleus of the host (but for some reason not the other way around), and the two ultimately became one. Happy story . . . but that's all it is. There is no explanation of a mechanism by which this could be accomplished. (Mechanism: Remember, every event [effect] must have a causal, a motive power. The "effect" of telling time via a clock is "caused" by the mechanism with. A poor analogy perhaps but easy to visualize.) This method (or something similar) *did* occur because we are told that it did.

The next step in the process is that these multicellular organisms formed themselves into colonies. They joined together for mutual

JOURNEY TO CLARITY

benefit. How these colonies gave up their individual rights and learned to reproduce as a whole is not understood, but we need not quibble over details. From this common ancestor began the branching out into the lines that would produce plants and animals. The lines that created algae and plants contained chlorophyll and mitochondria while those that wanted to become animals managed to divest themselves of their chlorophyll.

About this time, members of these colonies discovered sex. This isn't much discussed (especially in school texts) because, frankly, they don't know how it came about. A species (well, actually two members of the species) must pass only half of their chromosomes (meiosis) and the other pass only half of theirs (the correct half) at exactly the right place and the right time in evolutionary history in order for this to happen. There is no mechanism that has been identified that would allow this to occur. This is nothing short of a miracle—an odd claim by a group who categorically denies the possibility of the supernatural.

Nevertheless, our lines begin to diverge, and the first one we will follow is that which leads to plants. It is generally considered among evolutionists that plants evolved from algae that colonized the land. This may seem plausible at first until we look more closely at what must occur biologically in order for this to happen. Algae developed as an aquatic plant. Now, not to be argumentative, but I can't help wondering why it wanted to leave its tropical paradise and move onto the hot, dry shore. Did it bob around in the water gazing longingly at the strange new landscape? That aside, consider the changes it put itself through: it had learned to nourish itself underwater, keep itself vertical underwater, rooted itself underwater, and was protected from harmful light/heat underwater. It seems to me that as soon as it popped its little head out into a new environment, it would have died. To move onto the land, the algae would have had to collect new information in its chromosomes (without a pattern to go by). How, may I ask, does a plant *learn* new things? Granted, all things can adapt to an environment, but this is something else. This algae would have had to reinvent itself. An adaptation is something gained to *assist* an organism in surviving in a new environment (the color of a bear's fur, a shorter or longer beak for a bird, etc.). It would have to know how, prior to gaining the shore, to nourish itself from the dirt, root itself in the dirt, stay vertical in the dirt, and not only protect itself from heat/light but also thrive in it all while remaining an aquatic plant during the transition. That's one smart clump of green stuff.

It is said that to aid in the water-to-land transition, fungi came ashore in league with the new plants to aid them in drawing nutrition from the soil. This is called a symbiotic relationship (symbiosis), and we do see fungi and plants living in such a relationship today (in addition to many other examples). The inference here, however, is that they evolved *together* for this joint purpose.

Vascular plants (those plants with channels or ducts for carrying nutrition to the rest of the plant) evolved separately from algae. Suggested ancestors are various types of club moss, ferns, and horsetails (reproducing from spores rather than seeds). The carboniferous forests then appeared and are supposed to be the source of the coal and oil deposits scattered around the globe. Next up in the batting order were the seed plants. The first were probably those that produced seed via cones. These plants have seeds that are unprotected by a surrounding ovary or fruit referred to as gymnosperms. Likely candidates being conifers, gingkoes, and cycads. Flowers and fruits appeared next, which led to grasses, shrubs, and most trees (angiosperms).

Once again, there are problems galore with this steady march forward in the annals of life. One you've heard of before: where did the new information required for these changes come from? In addition, these symbiotic relationships are a two-edged sword. I'll explain: *symbiosis* is the term that applies to the various types of symbiotic relationships (scientists disagree on the number of them). Three of them are the following:

1. Mutualism: both organisms benefit from the relationship.
2. Commensalism: one organism benefits without harming the other.
3. Parasitism: one organism benefits to the harm of the other.

Some examples:

Mutualism: the shrimp digs a burrow where the shrimp and the goby fish live. Shrimp are blind, so the goby fish alerts it when predators are close by and the shrimp provides it a home.

Commensalism: Egrets eat the insects that are stirred up by cattle when they graze.
Parasitism: Aphids eat the sap from plants, depriving them of nutrition.

In the case of mutualism (which comprises most symbiotic relationships), it is assumed that the organisms and relationships developed concurrently. If they require each other to survive, and they did not develop concurrently, they'd be dead.

Another problem is that the gingko biloba and the wollemi pine are around today unchanged from those found in the fossil record. This would seem to show that evolution has both a fast and slow gear, so if the pace is unknown, how can any accurate measurement of time periods be made?

In any event, now that we have plants firmly established on the planet, we can move on to the advent of invertebrates (animals lacking a spinal column). How did all these diverse creatures come about? We mentioned previously that single-celled organisms gathered into colonies. These colonies were then able (the "how" being left unexplained) to form specialized cells and shaped themselves into a sort of bloated Pac-Man–looking creature that was able to ingest things. Eventually, one thing leading to another, we end up with lots and lots of different invertebrates frolicking about on the bonny blue seas.

As you have by now no doubt come to expect, there are significant problems with this notion as well. To become a multicellular organism from a colony of individual cells, the evolutionary process must step it up into high gear. New information must be obtained quickly in order to code the cells to behave differently and/or to perform new functions. It is not acceptable to say that the old cells will merely perform new functions. Without the additional information being precoded into their DNA, the old cells will just continue to perform the functions as directed by their original coding. In other words, one cell (call him Boss) must *learn* the functions of all the other cells and then *teach* it back to the new cells. The genes required for multicellular life must then be "switched on" and the now unnecessary ones "switched off."

When you read this in a school text, it is presented as a "just-so" statement. Stated another way, the authors present it as fact without any additional comment or explanation. It's like my saying, "Hello, my name is David, and I am the king of Norway"—period, end of sentence, on to the next subject. The "story" of evolution has been presented in just this fashion for at least the past fifty years in most of our schools and, despite the glaring lack of supportive evidence, continues to be so today. Comments like "We now believe," "a key moment in evolution," and "reaching a milestone" are inserted into the text without explaining *how* we know or *how* a key moment was reached or *how* the milestone came about. This is hardly *science*.

However, evolution waits for no man. Upon the horizon, a new age is dawning. While some invertebrates were happy campers remaining as they were, others yearned for adventure— an adventure on land. The vertebrates were born.

One thing, thus far, has not been emphasized enough and that is the subject of mutations. Just what are they in the evolutionary scheme of things? Descent with modification infers very gradual changes over vast periods. As we know, new information is required for something already existing to become something else, but in the case of mutations, information is generally *lost* during the process. Let me give you an example. As a fish grows legs, bit by bit, and crawls onto the land and becomes a crocodile it loses all the information that originally made it a fish. You can't go back and change the crocodile into a fish. That information is gone. The point is, on the one hand, evolution claims to be the information-gaining process while, on the other hand, the process that loses information. During the gain/loss process, the organism has to be able, every step of the way, to turn the old lights off and the new lights on (so to speak). It has to survive with a partial change in form.

Back in Darwin's day, nothing was known about DNA; but today is, as they say, a new day. As more and more evidence is discovered by *science* contrary to evolution, scientists continue to try to mold the theory to accommodate—like Silly Putty. The whole process is a litany of materialistic miracles.

Today, birds are the descendants of dinosaurs while it used to be taught that crocodiles were. Birds and mammals are supposed to have derived from different reptile ancestors. Feathers are placed on dinosaurs where none are found in the fossil record. Texts describe the very color and behavior of extinct animals that nobody has ever laid eyes on. "Feathered" theropods "discovered" in China were found to be frauds, but the corresponding news story was nowhere near as sensational as the announcement of the "incredible discovery." It was, indeed, literally incredible.

And so the story goes. The development of structures such as eyes and wings is inexplicable in evolutionary terms. Let's look at a common-sense approach to *why*.

Mutations are supposed to be improvements that gives the organism an advantage, assisting it in the struggle to survive. Those that are not effective in this regard are jettisoned. Let's take a fish as an example. We'll call him Fred. Fred has decided he would like to be an iguana, so he begins to spout little nubs from his bow and stern. (He knows he will need legs, so he's getting a start.) Now please tell me how this can be considered any sort of advantage for Fred? First, and most importantly, the girl fish will certainly give him a sideways look and not be willing for all the world to go out with him, rendering him impotent to pass on his nubs. In addition to that, he will probably now not be as efficient a swimmer because of the increased drag, his equilibrium will be disturbed, and he has more stuff sticking out that can become tangled in the weeds or snatched at by predators. Something else that really stumps me is why would Fred want to become an iguana? It seems to me that this would, in spite of Fred's iguana fantasy, result in a jettisoned mutation as one that is of no benefit if not downright harmful. To be of use, it must fulfill a function.

This brings us to another point: irreducible complexity.

Michael Behe is a well-known molecular biologist and the originator of "irreducible complexity." The idea goes something like this: the irreducibly complex is a system whereby the removal of any one part renders the system useless or, at the very least, unable to perform its intended function. The illustration he uses is the very common one of the old-fashioned mousetrap. The type mousetrap we're talking about is composed of five parts: a base or platform, the part that whacks the mouse, a spring, a bar that holds the whacker, and a catch to hold the bar. Now which of these parts can be removed and the mousetrap continue its function of trapping mice? None. The trap must exist in total, or it is worthless and will be thrown away. Descent with modification suggests that the, say, catch could exist independently and still serve a function. What function? This "mutation" would disappear as useless.

A mousetrap is one thing. Let's look at a biological example. In one of the many books about biology (or biology related) that I have read, I recall a great example of irreducible complexity regarding the human eye.

In order to have value in a population, mutations need to be of benefit so that they survive and are passed on. Now, with a system as important as an eye, we should be able to assume that it would be selectively of enough value to be inherited.

The problem would arise with the enormous number of mutations it would take to have all the necessary subsystems mutate at the *same time*. After all, with the human eye, we're considering more than just the commonly recognized parts like the pupil, cornea, and so on. We also have to account for the lens, the muscles attached to the lens, the retina with its 120 *million* rods and 7 *million* cones in each eye, and many other parts, not to mention the brain that has to know how to process the signals it's receiving and turn it into useful information. Every one of these component systems is made up of countless cells, each cell busily performing its own function.

Let's look at what goes on in the eyeball itself. Behind the eyelid, we find the sclera that encases the eyeball except at the very front where the cornea is located. The cornea is like a window that allows in light. The colored part of the eye is the iris, and it contains the black dot called the pupil. The pupil regulates the amount of light that enters the eye and adjusts automatically so that we can see in all sorts of lighting conditions. This is an example of the automatic controlling system that regulates our entire body (when it's in proper working order): the central nervous system. The left and right eye are connected to opposing divisions of the nervous system through attachment to two sets of muscle fibers in the iris. These muscle fibers radiate around the pupil like bicycle spokes.

We are able to adjust to lighting conditions because when light hits the retina, signals are sent to the muscle fibers, causing them and thereby the pupil to contract so that the light receptors within the eye are not damaged. If the light is dim, the muscle fibers contract the other direction, allowing more light to enter and to be received by the receptors.

Inside the eyeball itself, there are two chambers. A large one at the rear of the eye, about 75 percent of the space, contains a clear gel-like substance called vitreous humor (don't ask me why). Behind the cornea and in front of the eyes lens is a smaller compartment that is filled with aqueous humor, which is more like water in its composition. These fluids, in the proper amount, are what give the shape to the eye. The lens is held upright between these two substances by a set of muscles called the ciliary body. The retina lines most of the large compartment, which contains the vitreous humor. As light hits the eye, the cornea bends most of it, the lens completes the bending, and an upside-down and reversed image is passed to the retina. The retina and the sclera are separated by the choroid layer, which contains the blood vessels, which supply them with oxygen and nutrition.

The conciliary body connects to the lens via suspensory ligaments. When you are looking into the distance, the conciliary body relaxes, which allows the pressure exerted by the humors to cause the ring of muscles to expand, which causes the suspensory fibers to pull. This action causes the lens to stretch thinner, which minimizes the bending effect on any light passing through it. Naturally, viewing objects that are close cause the opposite effect.

The lens is constructed of cells that are interlocked and are long, narrow, and thin. As you would probably expect, these are highly specialized cells. These cells are filled with special proteins called crystallins that allow the lens to be transparent and flexible. The lens does not need any blood, which would make it murky in any event, but absorbs nutrition from the humors.

The cones and rods located in the retina are amazing things. Almost 75 percent of the body's sensory receptors are located in the eyes. It is via these cones and rods that light begins its transition from light into information. There are 120 million rods and 7 million cones in each eye. The rods function best in low light being specialized in monochrome and peripheral vision.

The cones detect colors and detail and operate best in bright light. Most of the cones are in the section of the retina called the macula, which collects the light when we are observing something directly. These cones come in three types: one picks up the color blue; one, green; and one, red.

When the cones combine their colors, we are able to see the world in "living color." The tips of the rods and cones stick into a layer of dark cells, and the other ends connect (synapse) with neurons that then synapse with ganglion cells (a mass of nerve cells). The connecting cells (axons) come together to form the optic nerve.

Cones and rods have both inner and outer segments. You can envision the outer segment as a stack of dimes enclosed by a membrane. This membrane has visual pigments embedded in it consisting of a protein and retinal. When light enters, it comes into contact with these discs (dimes), and the retinal molecule (which is bent) straightens itself out and detaches from its protein. This detachment changes the electrical charge to the membrane; passes the charge to a ganglion cell, which fires like a spark plug; and sends nerve signals to the brain. The visual pigments are able to rapidly (and I mean rapidly) reform ready to be used again. When discs become worn out, which they do on a daily basis, they are removed from the cone and rod tips and recycled by the pigment cells.

So far, we have light and electrical signals. How does that translate into sight? These signals are transmitted to the brain, and to prevent information overload, the processing begins in the retina. The 126 million rods and seven million cones in each eye send rudimentary signals, which must be funneled through one million ganglion cells that eliminate the weak signals. The stronger signals are forwarded through one million axons that are bundled together like wires in a conduit and form one of the optic nerves. There are two such nerves, one for each eye.

Just after the signal leaves the eye, the axons from the inner half of the retina cross over to either the right or left optic tract. The left tract carries signals from the eye's left retina and the right tract those of the right. Remember, the lens reversed the image, so the left tract is actually carrying the right side of the image and the right carrying the left side. Both tracts then enter a specific part of the thalamus. In this section of the thalamus, layers of specific neurons process the signals and interpret movement, color, or shape. Once the signals are interpreted and refined, they proceed on to the primary visual cortex.

In *Darwin's Black Box*, Behe cites twenty-one different steps that occur in the rods and cones alone and they take place in picoseconds.

A picosecond is the time it takes to travel the width of a human hair! Any single one of these steps missing, you are blind. Darwin and his peers had no conception of these sorts of processes. *Assembling the structural elements* in a mechanical system that will allow the chemical system to work.

The idea of irreducible complexity dictates that all the individual parts of the eye would have to have evolved at the same time and have been useful as independent parts. If they didn't, they would have provided no positive benefit and would have died out.

How does evolution respond to this?

First, secular biologists merely say that Behe and others like him are practicing pseudoscience. That's an interesting argument because it seems to me that the descriptions are either accurate or not. But they have something even better than the "liar, liar, pants on fire" rebuttal. It has been christened with the somewhat meaningless title of "punctuated equilibrium." Sounds impressive. What is it? In 1972, two paleontologists published a "landmark" paper developing their theory called punctuated equilibria. Stephen Jay Gould (1941–2002) and Niles Eldredge (1943–) postulated that once a species appears in the fossil record, they will become stable (stasis), showing little evolutionary change for most of their geological history. When changes do occur, they are generally restricted to *rare* and *geologically rapid* events where species branch out called *cladogenesis*. This process allows a species to branch into *two distinct species* rather than a gradual change over time.

WHAT? Talk about a convenient way to disregard irreducible complexity let alone millions-of-years evolution! In other words, time passes, a certain stage in gradual mutation is reached and . . . *BOOM!* Mother Nature says, "Let there be wings" and they appear! *WHAM!* "Let them have sight," and creatures now possess eyes! REALLY? Sounds awfully like miracle conjuring to me. Mr. Eldredge and Mr. Gould are/were educated to a much higher degree than I am. They have been lauded and awarded by academia, but they are die-hard soldiers in the war to prohibit any discussion (at least in scientific circles) of origins by any save material means. Their theory suits their world view. In their minds (I suppose), their theory *must* suit their world view. Their laudations and awards were largely given by folks who shared their world view. Indeed, practically to be *in* academia assures you of a warm reception. There is no better example of a materialist agenda that I can think of, and believe me, there are a bunch to choose from.

I am not going to spend any more time on evolutionary theory. Never did I intend an exhaustive commentary on this topic as there are already many available, and I am not trained to do so. What I *have* intended is to relate those features of evolution/materialism that had the greatest impact on *me*. In any event, it is my belief that any sincere searcher who has heretofore been impressed by evolution must surely, at this point, question its foundational principles. Evolution, Darwinism, materialism, theory of origins—whatever you choose to call it (it's all the same)—is a fatally flawed system.

But still, evolution rules our world. "So," you might say, "where's the harm in that? I mean, I see that it has more holes than Bonnie and Clyde's getaway car, but so what? Can't I live my life in peace without a theory of origins impacting myself and my family?" What started as Darwin's descent with modification via natural selection in 1859 has run tentacles through all segments of society. We'll discuss in the next chapter what some of the ramifications are, and then you can decide for yourself the answer to whether it matters or not.

Chapter Five

MY, HOW YOU'VE CHANGED

The influence of materialism—Epicurus, Lucretius, Darwin, as well as their modern sycophants—has had much wider ramifications than the accretion of planets and the origin of the eye, though with certainty their basis is founded on such. Based on this theory of how we came to be, direct inferences are made about how we should live our lives. If we are only the result of the indiscriminate collision of atoms, then what should direct our sense of family, right and wrong, good and evil? Do such concepts even exist?

If we desire anarchy, every individual may decide these questions on their own and act on those decisions. There is no objective reason for them to desist; rather, it would be the most natural thing in the world. In fact, to the materialist, there is no human reason at all. Reason is merely the particular firing order of our brain's spark plugs that keep us from falling off a cliff or wandering into a pond full of crocodiles. Anarchy, however, is a very precarious existence in general, so we have established societal rules of conduct, but they are not based on any objective morality. For example, if marriage between one man and one woman is deemed to benefit the overall good, then that is the rule. If, however, it is decreed by the most powerful (or loudest) citizens to be otherwise, it is dismissed and something, or nothing, else takes its place. Sexual fidelity is promoted for the stability of society, not to mention its general health, unless promiscuity is deemed preferable. If rearing healthy, well-adjusted children benefits the common good, it is welcomed; but otherwise, there is nothing inherently wrong with infanticide.

Many materialists will take exception to what I've just written, but it is the natural (no pun intended) conclusion that we must come to. We have no purpose, and our existence has no meaning. As accidental creatures, we possess no greater value than a cockroach. Some materialists may insist that "I love my wife and children," but what is love? It is merely the action of one seeking to promote the survival of his kind—the instinct to survive. The feeling or emotion he attaches to it is nothing more than the training he has been put through by generations that have sought to control him, or it is merely the echo of sexual excitement. It *cannot* be anything else if we are merely the product of chemicals brought about by chance. The bottom line is that we, the highest rung on the ladder of evolution (supposedly), have no purpose and will provide nothing that will have a lasting positive effect on the existence of anyone else. We are nothing more than bipeds who have come to consider themselves special because we *can consider* ourselves special. In short, we, for all generations past and to come, live in a constant state of delusion.

Epicurus and Lucretius taught men to pursue pleasure simply because they considered that it equated living a good life, one that held the most closely to reason. Now, Epicurus himself had very simple tastes and lived an ascetic lifestyle. He taught his students to try to satisfy themselves with small lightly sampled pleasures just as he did. However, the message was that there was no pleasure forbidden provided it was not taken to extremes because it would then become an obsession. Obsessions bring about their own type of pain, which is counterproductive. It takes a very disciplined man or woman to back off and control something for which they have acquired a taste. I have that problem with chocolate. There are many things, however, more insidious than chocolate. The underlying point here is that nobody can tell somebody else what is pleasurable to them. I might find pleasure in chocolate while you find pleasure in pulling the wings off flies. My disgust with that activity that you find so pleasing may be extreme, but I have no right to say you don't receive pleasure from pulling those wings off. How do I know? And remember, as a mere collection of atoms, it does not possess any immoral pretentions because there is no such thing as morality outside of what the collective says it is.

Lucretius enunciated materialist views further in his *De Rerum Natura* (*On the Nature of Things*). Here we find the evolutionary vision of society, morality, religion, and human nature clearly spelled out. To begin with, we find that the origin of the universe was nothing but the result of chance and had nothing to do with a designer or intelligence of any sort. The correct accidental alignment of atoms brought about the order discernable in nature. Lucretius repeats this mantra throughout his poem, thereby providing future generations with the notion that things can actually come to exist by *chance*. Lucretius says: "For certainly is was no design of the first-beginnings that led them to place themselves each in its own order with keen intelligence, nor assuredly did they make any bargain which motions each atom should produce; but . . . struck with blows and carried along by their own weight from infinite time until present, atoms have been accustomed to move and meet in all manner of ways, and to try all combinations, whatsoever they could produce by coming together, for this reason it comes to pass that being spread abroad through a vast time, by attempting every sort of combination and motion, at length those come together which . . . become the beginnings of great things, of earth and sea and sky and the generations of living things."

As further proof for the missing Godhead, Lucretius threw in the idea that the world is full of "faults." A god would never create a world where disease and wild animals would prey on man. Also, much of the earth brings forth nothing unless it is endlessly worked through human toil. (Oh, hateful work!) He provides us with evolution in ancient Latin form. His theory is that in the beginning, herbs and green grass covered the fields. The trees then appeared, producing their various kinds, followed by all the animals and lastly mankind. Of course, all this is presented in "just-so" fashion and not open to discussion—at least *serious* discussion. Human beings apparently had nothing to do with these faults, and here, I believe, the concept of "other-than-me" responsibility was firmly embedded in our psyche.

Lucretius lived in a time of tumultuous change though he was probably unaware of it. The world began to witness the spread of Christianity. We will not pursue the lines by which Christianity marched (and continues to do so) throughout the world. Suffice it to say that it would reach as far as Ethiopia, India, Russia, and the Mediterranean Basin and hold sway for over 1,500 years. Indeed, the apostle Thomas carried the faith all the way to India. To be sure, the rise of Islam caused the geography to change, and it continued to flow back and forth until settling into the alignments that we generally see today. Beyond question is the fact that strong roots were spread from Rome (the very despot that had tried so hard to destroy it) that reached throughout Europe and, eventually, to the New World.

As stated previously, during the fifteenth century, Europe was overtaken with a lust for all things ancient as a revival in learning was taking place. The so-called Dark Ages and medievalism had ended, and a period started that has been passed down to us under the label of the *Italian Renaissance.* About fifteen years after the discovery of the manuscripts of Lucretius, a full Latin version of Diogenes Laertius's (d. AD 240) *Lives of the Eminent Philosophers* appeared, which included a biography of Epicurus along with his three letters, which fully delineated his doctrines. Materialism was made available to a new audience, and that audience would increasingly become one that *read.*

By the time Darwin published his *Origin,* science was a freshly plowed field ready for the seed of a comprehensive theory from which to flower. Once the theory of descent with modification via natural selection was firmly established, natural laws quickly became inflexible and virtually omnipotent; the secondary causes had replaced the primary, or first cause, as the focus of science. Again, to cite our friend Thomas Huxley, he claimed that no loyal soldier of science could at the same time be a son of the church. It would be a contradiction. Edward Youmans (1821–1887) at about this same time hoped that he would live to see the day when science finally sounded the death knell of the superstitions (Christianity, voodoo, elves, etc., which were all the same), which had troubled man throughout human history (remember Epicurus!).

Youmans was the founder of *Popular Science Monthly* magazine in the first issue of which he wrote an article relating how science was now regarded as applying not to any single class of objects but to *all* of nature. He was convinced, and sought to convince others, that the operation of the universe could be explained by natural laws and the introduction into the process of any deity was met with open hostility. Indeed, he considered the concept of cosmic evolution an accomplished fact and wrote of it as such. Many others, of course, joined the chorus and have continued to do so down to our day. In the twenty-first century, it is assumed that everyone except the grotesquely ignorant or emotionally unstable understands the truth of the evolutionary process.

Before we proceed, let's look a little at the connection between Epicurus and Mr. Darwin. Darwin accepts on faith a materialist foundation. In addition, he merely exchanges the term *natural selection* for the magic of the atom. You can readily see the need for practically unlimited time as it relates to the accomplishing of either theory. This was a small matter when the universe was considered infinite. Both men considered morality to be subservient to prevailing social conditions and culture. In this regard, Darwin went further than the more timid ancients as he believed even human nature could be conquered. How? Through artificial selection—in other words, selective breeding.

Earlier, we mentioned that Epicurus indoctrinated his followers by exhorting them to teach themselves to think in a certain fashion; if they trained themselves to view the world in a certain way, it would be so (at least to them). Sort of like a group hallucination. So to with Charles, who admitted that his entire theory was hypothetical. Darwin went on to confidently argue that he didn't think any other explanation would nearly answer the questions contained in nature so completely and satisfactorily as his. He expectantly left it to future generations to *prove*.

It is interesting to analyze Darwin's *type* of argument. Throughout *Origins*, he claimed that the only theory that rivaled his own was that of independent creation. He presented each progression, or step, in his argument in such a way that the reader was forced to choose between believing in fairies, sprites, and Bigfoot or evolution. Put in this light, it appeared that only the uneducated would possibly conclude that the universe was the result of miracles rather than mechanics. Given the intellectual climate, Darwin was confident of a receptive audience. And so it is today. If you are not a subscriber to evolution propaganda monthly, you are an idiot or worse.

Darwin went further yet. Mention was made of the positive hate contained in religion as represented by witch trials and inquisitions. He was also quick to offer the suggestion that a grateful human race should laud science (and thereby Mr. Charles Darwin) for the improvement in reason and general knowledge—general knowledge and reason according to materialist principles, of course. In his *Descent of Man*, however, his final conclusions regarding humanity showed themselves in all their starkness and, ultimately, despair.

Sometime during all this deifying of science, German philosopher Freidrich Nietzsche could be heard crying, "God is dead." For those who listened, a still small voice could be heard calling back, which said, "Nietzsche is dead."

Today, many would have us separate the historical writings of Darwin from such modern nasty offspring as eugenics or social Darwinism. These are social outcomes that are, however, the natural-born children of his theory. Darwin had stated his conclusion that human nature was pliable, and with proper artificial selection, we could mold man and society into any form we wished. We merely had to screen citizens for the traits we wished to pass on. (Apparently, no one bothered with the question of just who would do the screening.) In his view, the view of science, morality was one more phenomenon to be explained by natural selection. Put another way, morality was nothing more than a side effect of chance.

Darwin's explanation in *Origins* went something like this: Intellect and morality developed through natural selection, and since natural selection varied due to climate and other environmental conditions, morality is, by definition, variable as well. We first adapted into societal beings because only on a basis of social organization could we live in relative safety. To keep order in this society, a sense of morality in the form of rules would be a necessary condition. In other words, an instinctual feeling must have been arrived at that convinced our unintellectual ancestors that it was better to live in groups than to live alone and living in groups necessitated some rules. And since strength was the great arbiter, I would imagine that these rules, in this early society, were promoted by the strongest in the group. The reason this feeling was arrived at is because of the incessant warfare they were supposedly engaged in. This caused the instinct for survival to kick in. It is revealing that his reason for the development of intellect and morality was necessitated by war. (I should think a psychiatrist would have a field day with this. Well, maybe not today.)

I believe a necessary condition for the presence of any sort of moral code is the possession of a conscience. Conscience is, I further believe, that spark of an objective moral code that all people are born with. As the conscience is something that is very difficult to deny, in order to fit it into the general scheme, it was also proposed that it resulted from natural selection just like everything else. Its evolution was explained as instincts that go unsatisfied; the condition of dissatisfaction comes about when we are possessed of two conflicting instincts. This argument goes on to say that we were possessed by societal instincts before personal ones (?) and we became conflicted when the two clashed. As a result of this clash between dueling instincts, conscience arose from the smoke and ashes of internal conflict. Of course, morality, being only a condition of necessity as we evolved, was but a social construct and possessed of no reality save that of expedience, so depending on whether you evolved in Ohio or Pakistan, your conscience and moral code would have evolved along different lines. Stoning women might be considered wrong in Dayton but perfectly fine in Islamabad.

The next interesting conclusion drawn from this theory on the development of intellect and morality has to do with folks that come from different cultures. As mentioned above, since your sense of morality would depend on where you evolved (became social), there would naturally be many variances in the concept of right and wrong, good and evil, etc. If you did not evolve is the *right* place, you would not have inherited the *right* characteristics necessary to advance the human race. (This would raise eyebrows later with the Darwinian "value" ranking of races and individuals.) "Correct" traits could be bred into the next generation of man. Exactly who would select these desirable traits was left unanswered.

This concept is the very foundation of what would become known as the eugenics movement, which started in earnest early in the twentieth century and was adopted as a central plank in the platform of Planned Parenthood. (Eugenics: the science of improving a human population by controlled breeding to increase the occurrence of desirable heritable characteristics.)

Darwin had considered at length whether the different races of men were, in fact, of different species. He stated in *Descent of Man* that if specimens of a Negro and a European were presented to a naturalist with no further information, then the naturalist would be forced to conclude that they were of different species. *Species* was a rather loose and arbitrary label, so Darwin felt at liberty to force rank the diverse races of man. Native Americans, Asians, and Negroes, it was stated, differ as much in intellect as in physical characteristics from the Western Caucasian. It was, therefore, permissible to place them on different rungs of the evolutionary ladder.

Darwin did not make these comments in passing. He remained optimistic about the progress of man, but he was emphatic about the importance of not allowing the *lesser sort* of men to outbreed the better classes. He did not propose this tongue in cheek. For instance, he claimed that the Irish lived in squalor and were largely unschooled (his thoughts, not mine), tended to breed like rabbits while the frugal, and a self-respecting Scot tended to marry late and have far fewer children. This sort of unencumbered, unchecked breeding, if allowed to continue, would certainly impede the progress of society. To prevent this catastrophe, he declares in *Descent* that human beings must take "natural" selection into our own hands. Further, he adds that we must take as much care of our breeding as we do that of our horses, cattle, and other livestock. (What a hopeless romantic.) People possessed of known defects, either physical or emotional, should refrain from marriage and of a certainty from childbearing. Darwin must have considered himself a fine specimen because he and his wife, Emma, produced ten children. I hope they looked like her.

It's small wonder that modern evolutionists tend to distance themselves from such sentiments. But why? Such beliefs follow in the train of materialism as naturally as light follows the rising sun. If we are nothing more than accidental collision of atoms, there is no more importance attached to a human being than that of an alley cat. If we consider the age of life as the determining factor, then bacteria are the most advanced life-form around because they have been here the longest. I have often thought bacteria deserving of more respect.

As we have discussed previously, the world view we select necessarily possesses moral implications. It is beyond question that it is in this area that materialism has had the most destructive influences on society. The many men and women who supported the advancement of this theory (and those who continue to support it) through the decades did not do so without understanding what they were doing. Many sociological theories have taken root as a direct result of evolutionary theory. Some of their authors include Karl Marx, Socialism/Communism; John Stuart Mill, utilitarianism, George Hegel, idealism; Adolf Hitler, national Socialism; and Friedrich Nietzsche, nihilism. All these men were Darwinists and had no respect in their various worlds for individual worth, the human soul, heaven, or hell. While we cannot investigate the ideas of all such persons, we will look at three of them.

Ernst Haeckel (1834–1919)

Ernst Haeckel fell in love with Darwinism upon reading *The Origin of Species*. He was irresistibly drawn to the unified concept of nature that it represented. Haeckel immediately understood the full ramifications of the theory and developed a materialist faith he called Monism. In his mind, evolution had made possible the ability to view all the universe in a mechanical, causal manner (omitting the First Cause, of course), the result being that the distinction between animate and inanimate bodies does not exist.

In his book *Wonders of Life*, he attempted to show that man was only a passing phenomenon, and he did this by comparing embryos of different species. He pointed to the similarities in all the samples examined as proof of descent from a common ancestor. His "proof," however, turned out to be his own drawings of embryos, which the scientific community (to their credit) forced him to admit were frauds: drawn by him with the specific intention of providing the "evidence" to prove his theory. There were those who applauded his efforts (apparently, the majority), however, and his drawings continued to appear in school texts for more than a century! His commitment to materialism justified perpetuating the hoax. The desired ends were attained, and there was no need to question the means used.

There was a more sinister side to his cunning. For Haeckel, the belief in an immortal soul caused the confused of the world to consider human life sacrosanct—a view he loathed. This idea, if allowed to continue, would permit the biologically "unfit" to continue reproducing. As claimed by Darwin, the unfit multiply like roaches, so the fit would become ever more scarce. There would be benefits to artificial selection (he termed it *marriage*) if it were properly directed. If society caused all those who were challenged physically or emotionally to be killed off, eventually the ship of mankind would be put right. In Haeckel's view (and that of other Monists), if there existed any such thing as morality, then the height of immorality would be to let such cursed creatures live.

Even were we to attempt to be gracious, it is not possible to overlook the horrid use his ideas were put to in Nazi Germany; the trail of the crime is too easily followed and documented. His popularity caused the population to make him one of the most admired authors in Germany. Arguments for eugenics were seized upon and accepted as sound. Hitler himself had knowledge of Haeckel's basic tenets and regurgitated them in language that was easily recognized. How a sane mind can absorb all the grotesqueness perpetrated under this regime of Haeckelian/Darwinian terror I cannot imagine. It is only with a broken heart that one can realize that the sick and twisted were left to slaughter the merely simple solely because they possessed the power to do so and the world view that led them to believe that it was right.

But Haeckel didn't stop there—no, sir, not him. A man committed is a man to be reckoned with. To him, a man was merely a biological lump; so he openly advocated for the legalization of suicide, euthanasia, and abortion—forcibly if necessary. It is important to realize that his ideas were not confined to the beasts of Nazi Germany. He sold hundreds of thousands of books in his home country, and they were translated into at least twenty-five languages.

Margaret Sanger (1879–1966)

Ms. Sanger is the founder of the multimillion-dollar ($229 million in 2011) international organization for the promotion of social manipulation: Planned Parenthood. While it is the largest provider of abortions in our country, it has also been at the center of altering concepts of morality since its inception in 1916. From the beginning, Sanger was convinced that the race of man that had been built through strength and power (natural selection) had long since been corrupted by such sentiments as compassion, pity, and tenderness. Let her own words speak for themselves: "In the early history of the race, so-called 'natural law' [read *natural selection*] reigned undisturbed. Under its pitiless and unsympathetic, iron rule, only the strongest, most courageous could live and become progenitors of the race. The weak died early or were killed. Today, however, civilization has brought sympathy, pity, tenderness and other lofty and worthy sentiments, which interfere with the law of natural selection. We are now in a state where our charities, our compassionate acts, our pensions, hospitals and even our drainage sanitary equipment all tend to keep alive the sickly and weak, who are allowed to propagate and in turn produce a race of degenerates."

Sanger

Planned Parenthood was originally a strong proponent of eugenics. (For all I know, they still are. I guess if they can sell baby parts, there isn't much they would consider unethical.) Sanger was an ardent supporter of the legalization of birth control, which in itself is not necessarily a bad thing, but she viewed it as a voluntary first step toward the eventual acceptance of a national eugenics program. An original slogan used by Planned Parenthood was "Birth Control—To Create a Race of Thoroughbreds."

Unlike Haeckel (who was considered sexually conservative), Sanger openly espoused the liberation from conservative (meaning self-controlled) sexual values. Sex was merely a biological act. However, with this freedom, careful attention must be paid to not allowing the mentally or physically defective to reproduce. She supposed that possibly draconian measures would have to be implemented in America should we fail to adequately monitor such breeding. This condition would undoubtedly present itself should we continue to act upon our stupid and cruel ideas of sentimentality. Sanger included in her notion of "mentally unfit" everyone from outright lunatics to those scoring in the low range of the newly implemented IQ tests, which would include, presumably, anyone who did not agree with her program. Sanger railed against the emotionally and physically crippled and the curse they presented to glowing humanity by their continued presence on this side of the sod. It was, to her, the height of cruelty to allow such misfits the opportunity of passing their defects on to another generation. In this regard, she was a beacon of open-mindedness; she considered male and female defects to be *equally* dangerous.

It should be noted that Ms. Sanger was good to her word and practiced what she preached. Sexual liberation was high on her list of goals and enjoyed at least an equal place with her desire to rid society of any unwanted pregnancies that resulted from said liberation. While she worked to advance these principles, she received help from one Havelock Ellis (1859–1939) who was a self-proclaimed "sexologist": Ellis leant her views on sex a veneer of "science." He wrote *Studies in the Psychology of Sex*, which sought to render the pursuit of sexual pleasure an end in itself (hail Epicurus!). Giving due credence to their joint beliefs, he and Sanger had a physical relationship while both were married, which stressed poor, unenlightened Mrs. Ellis to such an extent that she attempted suicide. Soon thereafter, Ms. Sanger divorced her husband, left off with Mr. Ellis (I suppose), and married one J. Noah Slee (founder of 3-in-One Oil) from whom she forced a prenuptial agreement that proclaimed their marriage to be open. This was a perfectly natural arrangement, given her beliefs, and sure enough she proceeded to have at least seven other affairs of the heart (or loins), one of which was with noted author H. G. Wells.

One final note on Ms. Sanger's contributions to society. She also had very definite ideas on how our children were to be educated so as to keep them from forming any potentially hostile repressions of their instincts. She recommended a program of sexual education, which started when children were five years of age and carried on until age eighteen when they should be well versed in "sexual expression"—in other words, active.

Alfred Kinsey

Yet another example of the legacy of evolution passed on to unsuspecting society is that provided by Alfred Kinsey (1894–1956). Mr. Kinsey wrote two books that were to lend scientific credence to the idea of sexual liberation. They were *Sexual Behavior in the Human Male* in 1948 and (you guessed it) *Sexual Behavior in the Human Female* in 1953.

Kinsey

It will become obvious that his distinction between *Male* and *Female* was clearly fuzzy. According to Kinsey, we are the result of sexual repressions absorbed from our culture and traditions and by ridding ourselves of said repressions, we explode (his word), allowing us to blossom into the full expression of our natural selves. Kinsey used charts and graphs to demonstrate his thesis and inserted scientific-looking documents as well. He was head of the seemingly innocuous Institute for Sexual Research, and within those hallowed halls such unrivaled depravity reigned as would make the Marquis de Sade blush. All the data used in producing Kinsey's findings was autobiographical—from personal experience. As the case would have it, the findings were geared to the ends desired.

The young Kinsey started out as a model student. His father was a Protestant of Calvinist convictions, who brought up young Alfred under a strict moral code. Alfred, however, in spite of his desire to conform to that code, was obsessed with sex and demonstrated homosexual tendencies. He engaged in masochism, which released the sexual frustration and the guilt he associated with it. His desire for self-control lost out to an overwhelming sadomasochism, which he pursued in earnest as he grew older but which he was very careful to keep hidden from the public. He was a very disturbed man.

Kinsey earned his doctorate in science (taxonomy) at Harvard. While studying there, he was influenced by William Wheeler who was a devout evolutionist and ardent adherent to the principles of eugenics. Kinsey became a fervent believer in the necessity of social engineering. But he ultimately left behind eugenics to pursue his first love: sex. In materialism (sex merely as a biological act), he could reconcile the conflicting mores of his youth and satisfy his sexual tendencies which by this time had wandered off into the realm of the bizarre. (At least, I find them bizarre.) To bring about his revolution socially, however, he realized that he would have to be discreet as there were still enough "moralists" in the world to be outraged by his doings.

Beneath the guise of science, inside the Institute for Sexual Research, a vision of Kinsey's new, sexually liberated society could be seen by any bold enough (or sick enough) to enter its inner "laboratory." Here he created what resulted in his very own pleasure palace. Staff members (and their wives) were encouraged to exchange partners and perform sex acts regardless of gender.

The outside world knew nothing at the time about the activities of Alfred Kinsey and his willing associates that occurred in the memorable halls of the Institute for Sexual Research. The fact is, if his deeds were known outside those walls, he is as likely to have been sent to prison as not. Instead, he was (and is still) proclaimed a giant in the gilded halls of science and was considered the founder of our sexual revolution (now there's a badge of honor). Indeed, there is an institution of higher learning named after him: The Kinsey Institute in Bloomington, Indiana. But more to the point for the subject at hand is that Kinsey assumed the concept of relative morality from Darwin and accepted that moral codes were only subject to the limitations society choose to place upon them. There is nothing *real* regarding the concepts of right and wrong when evolution is taken at face value.

To support his theories, he used examples of depravity as witnessed in society. These acts were not, in his view, episodes that should be looked upon with loathing and disgust but rather as natural reflections on how some chose to consummate their personal ideas of physical pleasure. He presented these occurrences as something that happened rather frequently and thereby, through their frequency, sought to make them appear normal. Society merely needed to be more open to, and accepting of, the wide diversity of human sexual behavior. Further, by not embracing these aberrant acts (my words), the victims (to me) were not to be pitied so much as the perpetrators for we were causing them, the perps (!), irreparable psychological harm.

Naturally, his manipulated *scientific* findings served to bear him out. In all his "studies," in which he used prisoners (which included those incarcerated for acts of pedophilia and bestiality), he specifically targeted those who were more likely (or, indeed, were known) to participate in nontraditional practices. When the figures did not correspond to the results he wanted to appear on his charts and graphs, he simply changed them so that they did.

Kinsey's attitude toward these two acts, pedophilia and bestiality, are perhaps the most disturbing of all—if one can make such distinctions. The source of Kinsey's data on many of these topics was one Mr. X. He was a middle-aged, quiet, and unassuming man who made startling revelations about himself. On all the earth (or under it), I'm not sure Kinsey could have found another man so completely devoid of scruple—other than himself, of course. He is said to have had sexual encounters with over 200 family members of both sexes and several animals. What fascinated Kinsey most was that Mr. X kept careful notes on these exploits including his partners reactions. One can only imagine what the reaction of the poodle was.

Much more could be said about our hero Mr. Kinsey and his "staff" at the Institute for Sexual Research, but we have said enough to make our point. The information I have left out is even more disturbing, beyond what I hope you can even imagine. Lest we consider Alfred Kinsey to be only an aberration, no less than Peter Singer, recent chair of bioethics at Princeton University, looks at bestiality as merely the last great taboo to be vanquished. For Singer, human-animal contact is as old as dirt, though he fails to mention where he has obtained his information. Whether this taboo is continued because of belief in humans as something special or some other sociocultural exclusion, it makes no sense in light of evolution. Singer claims to be truly compassionate to our fellow man, the taboo against engaging in physical acts with animals must fall—as long as we don't harm the animals!

Chapter Six

INTERMISSION

At this point in my journey, I took pause to consider just where I stood. The trip thus far had taken several years, and I had traversed many miles in the form of pages in books—many miles indeed. My mind was simply unable to accept evolution as a viable answer for understanding the how and why of life. But where did that leave me? Having only the supernatural and the non-supernatural world views to choose from and having rejected materialism, what was I left with? There are many scientists (more than are willing to make it public) who have come to accept a doctrine known as ID or intelligent design.

What is intelligent design? By and large, it seems to me, that ID seeks to respond to that aspect of evolution referred to as punctuated equilibrium by giving it a definite title. Or rather I should say by granting to it that which is obvious—at least to me—it has made of itself a kind of hybrid evolution/creation. Having discussed the preceding topics as we have, and knowing that we can choose between only two options, I would have imagined that design was obvious to anyone with an open mind looking at the evidence. But does ID provide us with any clarity? Does it promote the supernatural world view or otherwise direct us to creation via a special supreme being?

No, it does not. In fact, they are very careful to *not* make any claims regarding creation per se. Indeed, its website, IntelligentDesign.org, calls for a more developed evolutionary curriculum, which provides room for design as an option. So how does it help us on our journey? What assistance does it lend? None that I can make out. It leads us to the water but doesn't let us drink. Where does it think this evidence for design leads us? Who or what orchestrated that design?

Well, I couldn't buy that. Did they mean that evolution was real and some*thing* gave it a helping hand along the way? Was this merely a way to explain the weirdness of punctuated equilibrium? Surely, they would not suggest intelligent design without realizing that a designer is the consequence. I mean, after all, these are smart guys. Are they just afraid to go all the way and risk being totally rejected by their materialist peers? I'm sure that they, of all people, realize that they can be no more ostracized by materialists than they are by even suggesting such a thing as design in nature. Well, I can't account for them. Perhaps a world viewed through opaque glass is sufficient for them.

My next stop was in the East with the thought that maybe they had something I was unfamiliar with or hadn't thought of. There certainly is a variety of theology and philosophy to choose from there on the other side of the globe, and most of it, I admit, was new to me. The dominant ones are Hinduism, Buddhism, Jainism, Sikhism Taoism, Shinto, Confucianism, and East Asian Buddhism. When considering these, it is very difficult to separate theology from philosophy, and in some instances, it's impossible. There may be additional *ism*'s out there, but if there are, I found myself plumb ism'ed out, so I left them. I know that justice wasn't done to any of them, but they didn't seem to offer me any answers as most of them seem to be concerned with getting rid of this life with the least fuss possible. A little about each of them, however, follows:

Hinduism, as well as Buddhism, Jainism, and Sikhism, all originated on the Indian subcontinent and share several concepts such as dharma, karma, maya, and samsara.
Dharma: no one word in English defines it. It has to do with a principle of order in the cosmos as well as relating to Buddhism.

Karma: the totality of one's actions in this and any previous life, which determines their fate in future lives.

Maya: the power of gods and demons that allow them to create illusions and the power that allows the universe to delude us all in believing in its actual existence. (Gotta love that maya.)

Samara: the cycle of death and rebirth to which life in the material world is bound.

Hinduism is an ancient belief system (perhaps the oldest running in the world) and possesses voluminous writings of sacred teachings called Vedas. It runs the gamut from monotheism (through Brahman), polytheism, pantheism, monism, and even atheism. I admit to being confused and smelling of incense.

Buddhism is founded upon the teachings of Siddhartha Gautama who was himself the Buddha. Whether he actually looked like any of those statues you see about, I can't say. Never met the man. He possessed Four Noble Truths and the Eightfold Path and taught them to his followers. Buddhism is made up of the Theravada and Mahayana schools, and among academics, the Mahayana is further broken down into the East Asian form and Tibetan Buddhism. Buddha sought to help his followers to be liberated from samsara, the burden of going through life time after time trying to get it right.

Jainism are followers of Mahavira who claimed to be the twenty-fourth in the line of teachers of Jain. Jainist's reject the Vedas of Hinduism and practice a strict austerity, which will allow their souls to escape the cycle of samsara. Jainists also teach nonviolence, which covers verbal as well as physical violence (called ahimsa), and are, of course, vegetarians. The nonviolence applies to *all* living things. Not sure how they nonviolently harvest their vegetables, but I leave that inquiry to others.

Taoism (Daoism) are followers of the Three Jewels of the Tao, which are love, moderation, and humility. Generally, they seem to focus their theology on nonaction, humanism, relativism, and emptiness. It is a mostly polytheistic religion, but frankly the relativism and emptiness were more than I could handle, for it is these very concepts that I sought to rid myself of. It didn't make much sense to jump into the deep end of the pool here.

Shinto is a form of worship that focuses on family, tradition, nature, cleanliness, and ritual observance (like the tea ceremony).

Confucianism is a system of ethical and philosophical teachings founded by Confucius and further developed by a fellow named Mencius. (Mencius was best remembered for his stating that "human nature is good," making himself instantly, in my thinking, a deluded soul and not to be trusted.) Confucius taught love for humanity, value of education, ethical behavior in all quarters (including politics), ancestor worship, reverence for parents (one of my favorites), and harmony in both thought and conduct.

My apologies for any tongue in cheek that may have reared its head in the previous bullet points. Certainly, many people gain peace and contentment through the practice of the aforementioned, but they did not suit me. My answers were not there.

Let us shoulder our burden and continue on, faithful searchers to the end. When I had satisfied myself that science held no answers for me, I headed off into the land of religion. It became clear to me that Eastern belief systems held nothing relevant to my purpose. Most of those I considered did not concern themselves with a supreme anybody or the origin of anything. The focus of such religions seemed to me to concern themselves mostly with individual fulfillment that would allow one to meld into the eternal nothingness. Maybe I'm wrong, and if so, I hope those folks can forgive me. It's just too Krishna for me. I want something that explains where I came from and why I am here, and a lotus blossom simply won't do.

Now, admittedly, there are plenty of lunatic fringe beliefs in the West. We have been speaking of some of them during our review of evolution. After all, what difference does it make what you believe if you are nothing more than a highly evolved baboon? At about this point, I remembered Aristotle and that he was the guy that made the argument for a necessary First Cause. Where could I find out more about such a being? As a sidenote, in order to take a break from reading science—which, frankly, was getting old—I hopped on a bus and went into philosophy. Talk about getting lost in the tall grass! Just about when I was sorry I ever ventured there, I discovered Thomas Aquinas and he took me to the Bible. That's right—the Bible. The very book I had grown up with but had paid scant attention too.

Could that really offer any insight? I mean, Adam and Eve, Noah and the Flood (all I could see in my mind was all those cutouts of a Mr. Noah and Mrs. Noah who looked suspiciously like purported pictures of Mr. and Mrs. Santa Claus; cartoon giraffes poking the necks above a boat three inches long; and monkeys playing about on the deck), Samson, David, and Jesus? Really?

A couple things hit me right off. Adam and Eve—well, even evolutionists claimed we all descended from a single ancestor. I refused to buy into the idea that *all life* had sprung from a single cell that had arisen from nonliving stuff, but it just made sense that humans should have descended from humans. Also, I had discovered that the fossil layers took on just the form one would expect from a global flood. Besides, everything I needed to know was contained in one book instead of having to plow through hundreds as you have to do regarding evolution. It was time for me to put some serious study into this book.

Chapter Seven

CROSSING THE LINE

It might be wise, if you are easily hurt, to hold off telling people you know that you are going to begin consulting the Bible for reliable information. Especially should your friends be of an evolutionary bent. Hold off at least until the time that you have decided irrevocably on your world view as you will be considered as one who has gone off into the land of fairies, ghosts, and witches; and that can be hurtful if you are thin skinned. As an alternative to evolution, however, the Bible is the only reference that provides us with a coherent narrative that includes origins, the reason that the world is so messed up, human nature, and the grounds for hope.

You will be surprised, I'm sure, to learn that there are folks who have deliberately followed flawed dating procedures just to keep from using the Bible as a historical reference. Well, gasp in incredulity if you wish, but it is true. Maybe you thought that the search for truth was in some way the model for archeology? Sorry to be the one to tell you that not only is that not the case but you might stop expecting the arrival of the Easter bunny in April. Historians/archeologists are just as infected with the concepts of materialism as others.

Perhaps, the first question to be asked is "Why should I look to the Bible as a reliable source of information?" That's a great question and one that I asked on my own journey. After my study of evolution, with all the howlers contained therein, I eventually came to realize that a counterargument might be contained in *the* book. If a materialistic, natural universe didn't provide the answers, where else would I, after all, find the supernatural version?

The biblical record was written over a period of about 1,500 years. There are forty different authors, men from every conceivable background and vocation—from kings to shepherds. These men were either permanent residents or transients on three different continents and wrote three different languages. Contained within its covers, we find history, poetry, romance, law, parable, prophecy, and memoirs as well as other literary styles. Every known human misery is touched upon. All the failures of the character of men and women are displayed; nothing is left out or ignored. Interestingly, despite the diversity of these men, they wrote on common topics and with an amazing consistency of opinion. These failures of character (depending on your point of view) included marriage, divorce, remarriage, homosexuality, adultery, infanticide, authority, parenting, honesty, revelation and the very character of God. Where else can you find forty individuals, separated by time and distance, that would have universal agreement on such topics?

I found an interesting comparison regarding the unlikelihood of such agreement in a book I read a few years ago. It involves the literary collection known as *Great Books of the Western World*. This collection is comprised of over 450 selections from nearly one hundred authors, covering a period of some 2,400 years—some of the greatest, most respected thinkers in our part of the world. Philosophers, scientists, critics, theologians, writers of both fiction and nonfiction—all are included in this collection. Now, you'd think that being all Westerners you would find some underlying cohesiveness in their thinking, some regional commonality, some sense of, I don't know, brotherhood. Not much. If all these men were cats, you couldn't leave them in the same room together.

Something else I find interesting is the fact that in 2015, the United Bible Societies distributed over thirty-five million Bibles and twenty million New Testaments. That's just *one* distributor. The

Gideon organization put out an additional fifty-nine million! Over twenty million copies were *sold* in the United States alone! More than sixty-six thousand people are accessing a Bible application on a mobile device every *second*! In addition, the Bible has been translated into 2,200 languages. (I'll be happy if someone translates this book into English. That's supposed to be a joke.) After 150-plus years of Darwinism! I guess we, fellow travelers, are not alone in science not providing satisfactory answers to the truly important questions. The Bible has not only survived over time but it is also positively thriving.

There is, of course, a huge difference between possessing a Bible and looking to it as the foundation of a world view. Many Americans, for instance, include it in their collection of great literature, a Bible that is just another volumn gathering dust in a bookcase or on a shelf. I had an aunt once who had one prominently displayed on her coffee table. It was always there and only moved when she decided to dust the table. There are, indeed, many people in the West like my aunt, but it doesn't change the fact that the Bible is still the most reproduced, distributed, and reverenced book of all time. (The "best seller" lists don't count it as it is always number one, and they aren't interested in a perennial number one, especially when it's the Bible.)

In spite of the acclaim that the Bible has received, we are nonetheless forced to ask ourselves if it is to be relied on. Leaving aside the first ten chapters of Genesis for the moment, does the Bible prove to be historically accurate? If we should find that any persons, places, or dates should be proven to be incorrect, then we may rightly throw the whole thing into question. I say rightly because, ultimately, while we are looking here for much more than mere details, it is nonetheless in the details that the whole is revealed. The Bible claims to be the inspired, inerrant Word of God and therefore stakes a claim that is totally unlike any other book. It is, in fact, a remarkable claim, a universal claim, a life-changing claim (in every conceivable way). Can we, with rock-solid confidence, base our very lives on the truth of this claim? We shall see.

The Old Testament is the portion that we now turn our collective attention to. How, first, do we know that the Old Testament books that we have incorporated into our Bible are the same as the originals written thousands of years ago? Originally, the oldest important texts that we possessed were from about the third century BC. Of course, all these, and those of the next one-thousand-plus years, are in manuscript form. That begs the question: since they were copied by human hands, how can we know that what they say is what they said (as they say)? It's important to understand that the early Hebrews considered scripture transcription to be one of the most sacred occupations that could be entrusted to man. They were copying the very words of God. Remember that the Hebrews operated under a theocracy. In the state that had become Israel, to avoid any possible human error, a succession of scholars was devoted to the work of the "scribes." Originally, they were called the *sopherim* (which literally means *scribes*), and they produced their work between the fifth century BC and about AD 200. They were followed by the Talmudic period from roughly AD 100 to 500 and finally by the Masoretic tradition between AD 500 and 950. It's important to note the requirements that these scribes had to follow just to ensure their accuracy.

- The scribes must use the skins of clean animals only.
- They must be prepared for use in the synagogue by a Jew.
- The skins of the roll (scroll) must be joined together by strings from clean animals.
- Each panel, or skin, could only contain a particular number of columns, and they had to be equal throughout the scrolls.
- The column's prescribed length had to be between forty-eight and sixty lines and had to be thirty letters wide.
- Each section had to be first-lined (fine lines scratched into the skins so that the writing would be straight—like our modern ruled paper), and if three words had been written without a line, the section was worthless.
- Ink had to be black and made up by specific recipe.
- A certified manuscript was to be used to copy from, and no deviation was allowed no matter how seemingly insignificant.
- No letter or word could be written from the scribes' memory; the man had to look at each word to be sure that he was getting it correctly.
- A hair's breadth was to be left between each consonant.
- Three lines were to intervene separating the individual book.

- Deuteronomy had to end with exactly a full line. Of no other book was this required.
- The scribe was also to work in full traditional Jewish costume.
- Before beginning work, he was to fully bathe himself.
- He could not write the name of God after newly dipping his pen into the ink.
- If he were in the act of writing the name of God, he could not be interrupted even should the king himself come into the room.

Should it be observed or otherwise discovered that any of the rules were not followed, there were only two options for that particular piece of work: it would either be destroyed or sent to be used by schoolchildren in their classrooms. It could not be used in the synagogue. (I suspect that the scribe was also in for a counseling session with his supervisor.) Because of the exactitude in transcription and the Talmudist's certainty thereof, old or worn out scrolls were placed into a *gheniza* (closet) attached to the synagogue to be destroyed. This accounts for the lack of ancient copies. They simply were not concerned about inaccuracies.

When the temple was destroyed and the Jews dispersed in AD 70, the Jews, over time, began to assume the language of the various countries in which they settled. When the Masoretes came along (AD 500–950), it became especially important to preserve the Jewishness of the Old Testament. The text they took up from the sopherim was written only using consonants, omitting the vowels. The vowels weren't important simply because the way Jews learned to read and write taught them by memory where to insert them and how the words were to be pronounced. It was feared, after the dispersion, that the proper pronunciation would be lost, so the Masoretes inserted the vowels and punctuation. It is from them that we have the Hebrew Bible of today and, by way of inheritance, our Old Testament.

Lest we think that the fastidiousness of scribal work had become routine, the Masoretes put an additional set of regulations on top of those that already existed. They knew the number of times each letter of their alphabet occurred in each book. They identified the middle letter of their Bible. Every verse in every book was counted; every letter and unusual word was noted. It was through this seemingly excessive accounting of the scribal business, and the incredible awe of being allowed to write the Word of God, that we have handed down to us an accurate Old Testament.

In spite of all this meticulousness, the oldest text that we had of the Masoretic Old Testament dated from about AD 1000, and naysayers still harangued over the possibility of errors and laughed at it's being used for anything historical. That is until the discovery of what are called the Dead Sea Scrolls and improvements in technical aspects of archeology.

Beginning in the 1860s a German gentleman named Heinrich Schliemann—who had accumulated considerable wealth in the gold fields of California and traveling to Russia, selling military stores to the army during the Crimean War—retired from business and started to dig. He had a lifelong passion for the Greek classics and developed, at an early age (according to him), a desire to excavate the ancient city of Troy. At Hisarlik in modern Turkey, he found the site that he believed would ultimately reveal the ruins of Troy. Initial excavations did indeed expose an ancient city. Eventually, the outlines of walls appeared; and Heinrich was convinced he was on the site of Priam, Hector, Helen, and Paris, not to mention brave, bloodthirsty Achilles. Not being an expert, he immediately began to plow down, and thereby destroying who knows what, to the level at which he believed Homer's Troy lay. Long story short, he recovered some wonderful artifacts in gold and silver and christened them Priam's treasure. The point of the story is that archeology had revealed the reality of what had long been thought legend, and the digging started in earnest around the globe.

In the 1920s, the world was amazed and awed by the treasures taken from the tomb of the boy king Tutankhamun. We are still, to this day, dazzled at the discovery. It seemed incredible that so many items of beauty and worth had been buried with this fairly insignificant pharaoh (and without Bedouin weaseling their way in and robbing the tomb like they had so many others). This spurred on the interest and, as a side effect, caused the various governments to slam on the brakes regarding who had a right to what. Obviously, the home country felt they had the greater right as it was part of their national heritage and discovered on their turf, but the finders sometimes argued otherwise. Tourists, however, have always longed for souvenirs of their visits (preferably something authentic), so items on the black market or in small shops became a hot ticket. The poor in the surrounding countryside of likely sites knew this very well, which brings us back to the Dead Sea Scrolls.

A shepherd boy was scrambling among the rocks and boulders one day and tossed a stone down a hole. He was surprised to hear the crash of shattering pottery. Having managed to somehow squeeze himself between the rocks into what proved to be a cave, he found clay jugs lying about and stacked along the walls. Taking a few fragments of the writing he found along with him to his village, he scrambled back out and headed on home, hoping that his find would fetch him some coin for him and his family. Of course, questions were asked and the location of the finding revealed.

That was in 1947, and the Bedouin boy's name was Muhammad, and he had been looking for his lost goat. The hoard he had uncovered, which turned out to be thousands of ancient texts, had been lying in the dust and darkness of that cave for nearly 1,900 years. It had evidently been placed there for safekeeping in A.D. 68 during the then-current revolt of the Jews, which ended in the destruction of Jerusalem and its temple. (According to reports of the day, the Roman legions murdered over one million Jews on the spot and marched ninety-five thousand more into captivity. One of those taken captive was Josephus, the Jewish historian. The Christians in the city at that time are reported to have fled to Pella, a city seventeen miles south of the Sea of Galilee in modern Jordan.)

How did this cave full of old writings aid us in assuring a continued tradition of accurate transcription of Old Testament books? In the cave was found a full Masoretic copy of the book of Isaiah, which was dated from BC 125—a full one thousand years older than the oldest manuscript available at that time. It was found, according to Millar Burrows, that:

> Of the 166 *words* in Isaiah 53, there are only seventeen *letters* in question. Ten of these are simply a matter of spelling, which does not affect the sense. Four more letters are minor stylistic changes, such as conjunctions. The remaining three letters comprise the word "light," which is added in verse 11, and does not affect the meaning

greatly. Furthermore, this word is supported in the LXX and IQ Is (one of the Isaiah scrolls found in the Dead Sea caves). Thus, in one chapter of 166 words, there is only one word (three letters) in question after a thousand years of transmission—and this word does not significantly change the meaning of the passage. (Burrows, *The Dead Sea Scrolls*, p. 304)

I could go on regarding the reliability of today's Old Testament, but I won't as that is not our primary purpose. Except for one more point. Old Testament critics have long argued that the books contained therein could not have been written by the authors credited to have done so for the simple reason that the Hebrew language did not exist way back then. In 2008, Israeli archeologists discovered the earliest-known Hebrew writing. The site in which it was found overlooks the Elah Valley where David slew Goliath and was a forward Hebrew bastion in their ongoing skirmishing with the Philistines. The writing is dated between 1,000 and 975 BC—hundreds of years *before* critics claimed that the written language was formed. This is another huge biblical archeological find and a great segue to taking up a discussion of the historical accuracy of the Old Testament.

Like Schlieman's Troy, biblical place-names and persons were long considered the stuff of myth. If some event was known to have actually occurred, it was thought that the dates offered in the biblical text didn't match reality and/or were hundreds of years off. Admittedly, for many years the techniques did not exist for uncovering and deciphering the past even when the human will was interested in doing so. Since the 1920s, however, all this has changed. Merrill Unger (1909–1980) says: "Old Testament archeology has re-discovered whole nations, resurrected important peoples, and in a most astonishing manner filled in historical gaps, adding immeasurably to the knowledge of Biblical backgrounds" (Unger, *Archeology and the Old Testament*, 15).

Despite the preponderance of evidence, there are, as always, folk who want to obfuscate the findings. We are not going to bother here with arguments and are going to follow the majority opinion. Again, I'll only offer examples of the discoveries (or, more rightly, confirmations) as there are many wonderful books already available on the subject (included in sources and the end of this book).

There have been many astounding findings over the decades that have weighed the balance in favor of the historical accuracy of the Old Testament. For example, many critics have claimed that the religious practices of the Hebrews were sophisticated to the point that they must have been of comparatively recent origin—historically speaking, of course. When the city of Ebla was excavated with its over fifteen thousand tablets and fragments, much was revealed, which forced a revision of past prejudices. Kenneth Kitchen, in his book *The Bible in Its World*, said:

> In matters like priests, cult and offerings the records from Ebla so far merely reinforce for Syria-Palestine what we already know for Egypt, Mesopotamia and Anatolia in the third, second and first millennia B.C., and from the records from North-Syrian Qatna and Ugarit for the second millennium B.C. Namely, that well-organized temple cults, sacrifices, full rituals, etc. were a *constant* feature of ancient Near-Eastern religious life at *all* periods from prehistory down to Greco-Roman times. They have nothing to do with baseless theories of the nineteenth century A.D., whereby such features of religious life can only be a mark of "late sophistication," virtually forbidden to the Hebrews until after the Babylonian exile- alone of *all* the peoples of the ancient East. There is simply no rational basis that the quaint idea that the simple rites of Moses' tabernacle (cf. Leviticus) or of Solomon's temple, both well over 1,000 years later than the rituals practiced in half-a-dozen Eblaite temples, must be the idle invention of idealizing writers as late as the fifth century B.C.

Another bone of contention has centered on the date of the Hebrew Exodus from Egypt. The "official" date traditionally granted by historians is around 1230 BC. The Bible, however—in Judges 11:26, 1 Kings 6:1, and Acts 13:19–20—gives an earlier date of around 1400–1450 BC. Where does the "official" date come from?

It has long been taught that the protagonist Rameses mentioned in Exodus was Rameses the Great. This fellow was actually the second of that name. Little to nothing is known about the earlier Rameses

I, so the assumption was made that The Great had to be the one of Exodus. It should also be noted that the name Rameses was a common name in Egyptian culture and was also used to denote a geographic location or region. Further, it was known that there were no cities being built along the Nile Delta before 1300 BC. No cities being built, no need for oppressed Jews toiling away building them—thus, the earlier date and the discrepancy. Also, it was assumed that no Canaanite civilization was in existence at this time.

It is important to recognize that the history of the entire ancient world has been, heretofore, dated according to the record left us of the Egyptian kings. The reason for this was that it was always considered complete, accurate, and as recording consecutive successions of kings/dynasties. Two gentlemen, Immanuel Velikovsky and Courville, have shown that these lists of kings should *not* be considered consecutive or even particularly accurate. Some of the people listed were not kings at all but high government officials and some mere lieutenants, who may have been acting on behalf of the king, and others were known to have lived concurrently with the preceding dynasty. Viewed in this light, a new date for the Exodus can be given of 1450 BC or thereabouts. While not conclusive, it does allow one to no longer acquiesce meekly to a date of 1230 BC.

Let's take a minute to look more closely at these dating methods used by modern *experts*. There are two of them, and if I go into them too deeply, I will be the cause of putting all of you in a terribly confused state. Not wanting to make enemies of you, I will therefore just give you the baker's tour. Two systems are merged together to create the method whereby most archeologists chose to date people, places, and events in the ancient Near East and Egypt. The first is Sothic dating (falsely called astronomical dating). This theory of dating is not supported by any evidence and also flies in the very face of common sense. One feature of it is the assumption that those learned Egyptians that managed to build the pyramids did not know how long a year is. The position taken is that they used a calendar of 365 days as far back as 2500 BC or even earlier. Our year is actually 365 and 1/4 days; thus, we use leap year to adjust our calendar. Well, the stupid Egyptians didn't know any of this, so they just let their calendar get out of whack by about 1,460 years by going backward 1 day every 4 years (not accounting for the 1/4 day). So New Year's

Day would only appear on their calendar on the same date every 1,460 years. This 1,460 years is referred to as a Sothic period. (No other nation, old or young, known to have existed under the sun ever allowed their calendar to carry on like this.)

How was this arrived at, and what the heck does it mean? It is believed that Sothis referred to a star, in this case supposedly Sirius, that rises *at the same* time as the sun in the east. Well, first off I defy anyone to look east into the rising sun with their naked eye and be able to tell if there is a star anywhere near it. But this whole theory was devised based on a single fragment of ancient document that mentioned *Sothis rising*. Now, mind you, it is only one reference, and there is absolutely no mention of what Sothic/s is or refers to! It comes from a paper addressed to a priest named Pepihotep and is dated in the twelfth dynasty (the reign of Sesostris III). It says:

> You ought to know that the rising of Sothis takes place on the 16th of the 8th month. Announce it to the priests of the town of Sekem-Usertesen and of Anubis on the mountain and of Suchos . . . and have this letter filed in the temple record." (M-SEC, p. 46f)

This inscription has led to a convoluted, surmise strewn system of dating that would have appalled the very people it's attributed to. Its use in providing BC dates was not even proposed until the nineteenth century AD and then by one Karl Lepsius, a German Egyptologist. A gentleman named Meyer then came along and invented the calculations that were used to extrapolate the dates backward. It was determined that the star Sirius rose with the sun on the Egyptian New Year in AD 140. Working his magic in reverse, the year 4240 BC became the only date that could be *fixed* in ancient history. Unfortunately, an astronomer named Poole (whose findings were confirmed by the Astronomer Royal of London) did the astronomical calculations and found the Sirius did *not* appear on the horizon on that day! This finding, however, apparently didn't faze modern archeologists or alter their ideas.

This idiotic concept is used in conjunction with the list of kings provided to us by one Manetho. I should say *lists* as there are two. This gentleman was a priest of Egypt who lived during the third century

BC. At the time, Egypt was ruled by Greece because the armies of Alexander had recently subdued them. Mr. Manetho, in his desire to show that not only was Egypt greater that Greece but older, then drew up his great long lists. (Egypt was not, however, greater than Greece on the battlefield it would seem.)

Problems with his lists immediately arise. First, the two don't agree with each other! That aside, they are very long and include in them men who were not kings but viziers and subrulers, and there are duplicate listings of the same men. Some names may even be invented, though, to do justice to Manetho. we can't know with certainty. (Some of us are more fastidious regarding fact than others.) The inclusion of nonkings and duplications does lead one to the conclusion that, perhaps, Manetho was looking to create an effect. Rather than placing the kings in the order that can be supported factually and ridding the lists of repeat entries, modern archeologists take them—contradictions, errors, and all—as gospel. What is the reason for this? In spite of their own discoveries, they refuse to accept that the Bible can be in any way regarded as a reliable, historical document.

Manetho, the Egyptian priest, was believed to hail from Sebennytos and lived in or around the third century BC. His name has been passed down to us is Greek. His original Egyptian name is unknown, though, as is common in Egyptology many guesses have been put forth. He is remembered largely for his *Aegyptiaca* (*History of Egypt*), which includes his list of kings (or, rather, one of his lists) though he wrote other things including something entitled *Against Herodotus*. His birth and death dates are unknown, but it is suggested that he worked during the reigns of Ptolemy Soter I (323–283), Ptolemy Philadelphos II (285–243), and maybe even Ptolemy Euergetes III (246–222). If so, he lived a prodigiously long span, and I'm sure his kinsmen were glad when he finally skipped off. (It seems funny to me that dating Near Eastern ancient history can be so inexact at these relatively recent dates yet archeologists can be so cocksure when they speak of millennia earlier).

It was Manetho who first started to group kings into "dynasties," and he wrote all his works in Greek. His major in college seems to have been in learning the service of the sun god Ra with a minor in all things Sarapis (a sort of hybrid as the Greeks had crossbred Osiris and Apis sort of like a freshwater striper or a mule). His dynasties were generally divided not by bloodlines (as with other royal lines, say, in Europe) but by the geographical shifting of influence—the location of capital cities or sites of particular religious significance. Then again, he could shift into genealogical gear and lists kings as "son," meaning that this king was the begotten of the previous one. It has been said that he wrote his history as a rebuttal to Herodotus, which writer had previously gone to press with his *Histories of Herodotus*—a history of the Greeks. There may be some credence to the theory of Manetho wanting to one-up his contemporary because, as mentioned previously, he did write *Against Herodotus* though the original is lost to us. Perhaps his intent was to accomplish this by presenting something much longer and more ancient.

Many problems exist with Manetho's lists despite modern archeologist's reliance on them and insistence on their use. The lists no longer exist in total, and it has further been suggested that information found in Lysimmakhos was inserted into the original around the first century. The Jewish historian Josephus was the first to mention Manetho during his dispute over yet another list known as the Apion list. The Apion list did not include the eighteenth and nineteenth dynasties. Josephus tells us that Manetho admitted to using "myths and legends" and "oral tradition" in assembling his tables of kings. In fairness, we cannot ask Josephus or Manetho for their sources. They are dead, I think.

Further complications arise with the finding of additional lists. One similar to Manetho is the Turin Papyrus. We also have the Palermo Stone, which comprises what knowledge we have of the Old Kingdom rulers. There are other such stones with lists in Karnak and Abydos, which has two. These stones have inscribed on them names of New Kingdom kings. The differences between these stones and Manetho are great. For example, the Palermo Stone lists kings of both Upper and Lower Egypt. Egypt was divided in predynastic times. Manetho also includes names known to be those of Egyptian gods and not real men at all.

The stones at Karnak and Abydos have left off the names of Hyksos kings as well as any associated with the "heresy" of Akhenaten. It lists seventy-six names in total between the first and nineteenth dynasties. Yet another list, known as that of Saqqara, has only fifty-eight names and is believed to have been developed in the time of Ramesses II. As Manetho came along many centuries later, we cannot know whether he consulted these lists (or caused them to be buried under the sand). To make matters worse, from the fourth dynasty on, kings could and did have as many as five names—or more. (The earlier dynasties had multiple names but not as many as five. They were happy with two or three.) Now, I am no Egyptologist, but I have to ask: with all this uncertainty and confusion, how can archeologists pretend to be sure of much of anything?

(NOTE: **Hyksos**: Canaanites had moved into the Nile Delta around 1800 BC. They were followed by the Hyksos [about one hundred years later] who were folks that hailed from somewhere in western Asia. The fourteenth dynasty was actually a Canaanite realm that had established itself and which coexisted with the Egyptian thirteenth dynasty. Around 1650 BC, the Hyksos had become powerful enough [due to famine and other misfortunes] to subdue both of these kingdoms and, in their turn, started the fifteenth dynasty. It is believed by some that the duration of Egyptian rule can be reduced by 600 to 700 years by ridding the king lists of overlapping, duplicates, and gods.

Akhenaten's heresy: Egyptians traditionally worshiped a ton of gods. There wasn't room in an Egyptian road to walk for all the gods roaming about. Akhenaten worshiped one god named Aten (a sort of solar deity) and sought to change all his subjects to his way of thinking. This confounded many people, not the least of which were the thousands of caretakers of all the displaced gods who found themselves out of work. It didn't take. Akhenaten had his way while he lived, but no sooner was he cold than everything reverted to what it had been before. Poor Akhenaten had his name removed from the list of kings for his trouble. I hope he wasn't terribly troubled by this as the lists appear to have been somewhat capricious anyway.

It seems fair to me that when considering Manetho and the others who left us these lists, we can thank them for taking the trouble of leaving us the names but curse them for the consternation resulting from all the mess. While we can appreciate them, we should not consider them historic. At least I have a higher regard for history than to consider them as such. In addition, it seems the very definition of folly to disregard, out of hand, a complete historical record as presented in the Bible and exchange it for lists of kings that are full of arbitrariness, contradiction, and outright error. Lists that were drawn late (in the case of Manetho) from incomplete data and, as far as we know, at the current date, lists assembled with an agenda and with prejudice.

Why do I harp on all these lists and dates? It is because by adhering to them in all their inaccurate glory, it places the date of the Israelite Exodus from Egypt around 1235 BC while the Bible text gives it as around 1450–1400 B.C. The 1235 date does not fit the rest of the biblical text, so it lets modernist archeologists claim it all to be rubbish. According to current "expert" *opinion*, the 1450 date cannot be reconciled. Let us non-experts see, however, whether it can and not because we want it to but because it fits the evidence.

The collective evidence fits perfectly with the biblical record if we redate the Early Bronze period and get rid of the fluff and error in Manetho's lists. Can we reasonably do this? Are we altering things just to fit our world view (if it *is* your world view)? The clear majority of the argument today has to do with the unwillingness of modern archeologists (or so it seems to me) to adjust their thinking to their own finds. If we accept the large amount of historical and archeological data available to us today in their proper sequence, the arguments largely disappear. It makes one wonder why they are so unwilling to budge.

What changes need to be made? I believe that I have adequately covered the problems with Manetho's lists. Dynasty six and dynasty twelve are listed separately and hundreds of years apart, but they are *one and the same*. These aren't two dynasties but one. The events ranging between the enslavement of Joseph and the eventual Exodus are all supposed to have occurred in this redundant twelfth dynasty.

This reordering of dates is necessary to make sense of what we *do* know. Having put things in order, let's take a look at some other instances where events that were formerly contentious now line up using this revised chronology. (Well, to be honest, they are still contentious to materialist archeologists. Well, to materialists in general.)

There are several reasons for these proposed realignments in the chronologies as currently accepted.

First, there are several known concurrent people and events as recorded in both Egyptian and Hebrew records. Based on these records, the twenty-fourth to twenty-sixth dynasties should be placed after the fall of Israel to the Assyrians in 721 BC.

Second, if Joshua's conquest of the promised land coincides with a shift in the Early Bronze Age IV timeline, because these "ages" are directly tied to the chronology based on the lists of kings, the Hittite kingdom came to an end shortly after Rameses II rule and began with the Hebrew conquest. The reign of Rameses II must have ended about seven hundred years after the Hebrew invasion. This lines up with the approximate point of the fall of Israel to the Assyrians. Long story short, dynasty nineteen should come just before dynasty twenty-four.

Third, since dynasties eighteen and nineteen span about 350 years, they must have started about the time of the founding of the Israelite kingdom under Saul. The time prior to the eighteenth dynasty was the period of Hyksos rule. This would have been concurrent with the period in which the Judges went about the land of Israel.

Fourth, the only way this could have happened is if the twelfth and thirteenth dynasties ran concurrently with the sixth. The Middle Kingdom, as such, did not exist but was created in modern times based on Manetho's erratic lists. Only the Hyksos era separated the Old Kingdom from the New Kingdom, thereby eliminating the seventh to tenth and fourteenth and seventeenth dynasties.

Fifth, dynasty two was probably roughly at the time of the building of the pyramids. The correct list of dynasties should look more like first, fourth, fifth, twelfth, the Hyksos, eighteenth, nineteenth, and fourteenth to twenty-sixth. All the other dynasties are intermixed in some fashion with these. There is a detailed analysis of this and other topics in a book named *Making Archaeology Biblical* by Vance Farrell. I highly recommend it if you are interested in a more detailed, thoughtful study.

Take the story of Joseph and his many-colored coat, his jealous brothers, and his transportation to Egypt via a slave trader. Did it really happen? Was there a Joseph who, from slave, rose to be second only to Pharaoh in ruling Egypt? Let's see. Sesostris I (a.k.a. Senusret I and Senwosret I) is the logical (both in dating and through archeological finds) king who took Joseph from prison and raised him to the position of vizier. This particular vizier was given far more sweeping powers than any other mentioned in Egyptian records:

> Then Pharaoh said to Joseph, "Inasmuch as God has shown you all this, there is no one as discerning and wise as you. You shall be over my house, and all my people shall be ruled according to you word; only in regard to the throne will I be greater than you." And Pharaoh said to Joseph, "See, I have set you over all the land of Egypt."
>
> Then Pharaoh took his signet ring off his hand and put it on Joseph's hand; and he clothed him in garments of fine linen and put a gold chain around his neck. And he had him ride in the second chariot which he had; and thy cried out before him, "Bow the knee!" So he set him over all the land of Egypt. Pharaoh also said to Joseph, "I am Pharaoh, and without your consent no man may lift his hand or foot in all the land of Egypt." And Pharaoh called Joseph's name Zaphnath-Paaneah. And he gave him as a wife Asenath, the daughter of Poti-Pherah priest of On. So Joseph went out over all the land of Egypt." (Gen. 41:42–45)

Mentuhotep was the only vizier known to have such a unique position and authority in Egyptian history. The biblical name given by Moses as Zaphnath-Paaneah (Joseph) caused modern scholars a lot of trouble over the years, but we must remember that Moses knew a lot more about the Egyptian vernacular of the time than folks today. He was raised for 40 years as an Egyptian prince. I am not going to review all the controversy here for two reasons: (1) I'm not a philologist and (2) I don't care to (and I'm the one writing this!). The explanation that I have seen that makes the most sense is that the first part of the name in not a proper name at all but a unique title that translates into English as "overseer and/or minister of the storehouse of abundance." The second half of the name means something akin to "preserver of life." Archeologist James H. Breasted (1865–1935) gave the following titles to Mentuhotep based on the many inscriptions found: Vizier, Chief Judge, Overseer of the Double Granary, Chief Treasurer, Governor of the Royal Castle, Wearer of the Royal Seal, Chief of all the Works of the King, Hereditary Prince,

Pilot of the People, Giver of Good-Sustaining Alive the People, Count, Sole Companion and Favorite of the King.

There is an interesting aspect associated with these titles—that of Hereditary Prince. This has a direct connection to the reason for the Exodus about 230 years later. Being a prince, in addition to all his other responsibilities, Joseph also had to rule a specific province, known to the Egyptians as a *nome*. This would cause problems for the Israelites in future generations. Indeed, he appears in the Turin list of kings (twelfth dynasty) in the Egyptianized form of Yufni (or Yufsi). Joseph is also mentioned as being involved in an irrigation project, which sought to gather Nile floodwaters into catchbasins in order to increase the tillable land in the area that it served. This system of canals and ditches is not only known as the Canal of Joseph but is so listed on maps to this day.

Now, to get back to the impact of being a Hereditary Prince. During the time of the seven years of famine, the Egyptian people had to buy the grain from Pharaoh. During the first year, it cost them money. The second year they paid with their cattle or other livestock. Over the next five years the people paid with their land so that by the end of the famine, Pharaoh owned all of Egypt. Up to this time, the country was broken into *nomes*, which each had their own ruler who was left to his own devices as long as Pharaoh got his cut. Basically, Egypt was much the same politically, pre-famine, as feudal Europe. During this time of famine, royal princes and their nomes were exempt from paying for the grain and received a royal allotment gratis. Joseph's nome was most likely the land of Goshen where his kinsmen were settled. While the rest of Egypt grew poor, the Israelites flourished. They still owned their land, homes, crops, and cattle. Naturally, when the famine ended and the native Egyptians looked around and saw their impoverished condition in contrast to that of the immigrant Israelites, envy and even hatred grew into a powerful prejudice. Then there came a king "who knew not Joseph" and a time for get-even ensued.

We are told that Joseph was 110 years old when he died. It is probable, therefore, that Sostris III is the pharaoh that started the oppression of the Hebrews. We know that considerable changes came about under him that altered governmental policies that had been in effect for a least one hundred years. Sesostris III got rid of the feudal system that had been in effect and set himself up as a dictator. Sesostris III denuded the former nome rulers not only of their power but also of any material possessions beyond those he wished that they should retain. Surely, as the land of Goshen had probably become the fattest nome, it was not immune from this change in circumstances.

Further, given the location of Goshen connecting Egypt to the Fertile Crescent, any potential enemy of the state would probably invade through that territory, and it was feared that the vast horde of Israelites would join in the fun of pillaging the remainder of the country. The Hebrews were stripped of their special position, their land, cattle, and so forth and within a short time were further reduced to the status of slaves. It is known that Sesostris III and/or his successor Amenemhet III initiated a very ambitious building program, which used extensive slave labor. Unlike their predecessors, they used the medium of sun-dried brick to accomplish their designs. Josephus tells us that the Israelites also built pyramids, and in the twelfth dynasty, all the pyramids but one are of brick. The building efforts were concentrated in the very land of Goshen located in the eastern delta of the Nile. This area contains the sites of Pi-Rameses and Pi-Thom mentioned in the Bible. (There has even been archeological evidence of some of these bricks being made without the use of straw.)

But what about the Bible saying that the name of Pharaoh during the bitter toil was named Rameses? In the Sothis list of kings, just after the name of Uses or Unas, follow a whole herd of Ramessides after which follows the name of Koncharis. This string of Ramessides are identified as the kings of the twelfth dynasty and are alternative names to the other lists. In other words, the Rameses of the twelfth dynasty ran concurrently with the kings of the sixth dynasty. As mentioned previously, Rameses was a very popular name. This name was used as a personal name as well as for geographical locations. We also know that Egyptian kings enjoyed possessing a great number of names. The longer the list, the greater the man, I suppose. In addition, Josephus tells us in his *Antiquities of the Jews* (book 2, chapter 9, par. 1) that after the death of Sesostris II, the line changed. Sesostris III was *not* the son of Sesostris II as modern archeologists have long assumed.

Manetho tells us that the capital of the sixth dynasty was in Memphis though most of their activities were concentrated to the south. Dynasty twelve started with its capital in Thebes but early in its reign moved it to Ithtowe, just a few miles south of the delta. We know this by inscriptions left on monuments. We also know that the Egyptian kings had become fearful at the prodigious rate at which the Hebrews multiplied. It would be reasonable that this growth in population would have expanded to the south and west. Considered in light of the whole, might it not also be reasonable to conclude that the capital was moved for the very reason of being better able to keep an eye on these miscreants and to oppress them more effectively?

What about the story of Moses? Aside from his having an Egyptian name, do we have any evidence for him? Amenemhet IV ruled but nine years when he was succeeded by a queen who had the name of Sobekneferure. Apparently, there was a dearth of male heirs at the time. It might be the case that Sobekneferure was a coregent while the young Amenemhet IV grew to maturity and managed to outlive him. At any rate, this gal was already a "seasoned" citizen when she came to the throne. Interestingly, the tomb of Amenemhet IV has never been found though every other tomb of the kings of this dynasty have been located. Interested parties have searched high and low but no tomb. This leads to an intriguing theory, which I happen to like because it fits with the revised chronology (not to mention the evidence).

It could be that the said senior citizen Sobekneferure was the very Pharaoh's daughter that fished baby Moses out of the Nile. After years at court as her son, he takes off into the desert because he has killed an Egyptian. As far as the chronology goes, the foster mother seems to have married Chenephres who was a prince of the thirteenth dynasty. Is it possible that Amenemhet IV of the undiscovered tomb and Moses who had been raised as a coregent with Sobeknefrure for the nine years before he had to flee, were one and the same? The end of this coregency comes remarkably close to the time Moses split and gives additional meaning to the words of the apostle Paul:

> By faith Moses, when he was come to years, refused to be called the son of Pharaoh's daughter; choosing rather to suffer affliction with the people of God, than to enjoy the pleasures of sin for a season. (Heb. 11:24–25)

Exodus from Egypt

As Joshua entered the Promised Land, he and the tribes of Israel came upon the city of Jericho. Even if not a Bible person, you probably know the story of the fall of Jericho. Forget dating for the moment, and let's look at what the archeologists have actually found at the site. The pottery fragments have been affirmed to be of the Middle Bronze period (1950–1550 BC) by two gentlemen named Ernst Sellin and Carl Watzinger who excavated at the site. In addition, the wall foundations were at such an angle as would be the result of the their falling outward, away from the city interior. Evidence of burning is plainly visible on the remains of the cities' interior. All through Palestine, in every city and village so far excavated, the same evidence of destruction is found just as you would expect to find from an invading people intent to take over the land for themselves.

Dame Kathleen Kenyon (1906–1978) excavated at Jericho in 1952. Certainly, no friend of the historicity of the Bible, she did, however, albeit unintentionally, confirm the very record she sought to refute:

> The final end of the Early Bronze age civilization came with catastrophic completeness. The last of the Early Bronze Age walls of Jericho was built in a great hurry, using old and broken bricks, and was probably not completed when it was destroyed by fire. Little or none of the town inside the walls has survived subsequent denudation, but it was probably completely destroyed for all the finds show an absolute break, and that a new people took the place of the earlier inhabitants. Every town in Palestine that has so far been investigated shows this same break. (K. K., *Archeology in the Holy Land*, p. 134)

The old and broken bricks that were used to shore up the walls indicate that the citizens of Jericho were in an awful hurry to strengthen their defenses. Why? Because they had heard of the horde of Israelites coming their way and what they had already done to the Amorites and probably the Midianites as well. How do we know that the invaders were Israelites? Well, it is well-known that the twelve patriarchal families divided into the twelve tribes of Israel, presumably with some differences in social, familial customs. Dame

Kathleen also found evidence of this when it was discovered that there were burial grounds that displayed differing practices:

> The most satisfactory explanation seems to be that the difference in burial customs is due to a tribal organization. . . . It is interesting to find that at two other sites, Tell Ajjul [Gaza] in the south and Megiddo in the north, there is evidence of a similar organization. (K. K., *AHL*, pp. 141,143)

Another interesting find supports the biblical injunction against rebuilding Jericho. Joshua told the Israelites (Jos 6:26 NKJV): "Then Joshua charged them at that time, saying, 'Cursed be the man before the Lord who rises up and builds this city Jericho; he shall lay its foundation with his firstborn, and with his youngest he shall set up its gates.'" (Jericho must have been *very* bad.) Once again, the findings at Jericho verify the record. Dame Kathleen says:

> As has already been described, the latest of the Early Bronze Age town walls at Jericho was destroyed by fire. With this destruction, town life there came to an end for a space of several hundred years (*sic*). Newcomers, who were presumably the authors of the destruction, settled in considerable numbers in the area, but they did not build for themselves a walled town. Similarly, on the adjacent hillside, occupation debris and pottery is found, but no structures. . . . They destroyed existing towns, but did not create their own. (K. K., *AHL*, p.137)

It's easy to understand why all this garbage (what constituted garbage in the fifteenth century BC I can't imagine) accumulated in the environs of Jericho. Israelites were encamped for a considerable distance around the site for some years. The Reubenites, Gadites, and half of the tribe of Manasseh had already received their land on the east side of the Jordan River, but all the rest of the folks were on the west side. Joshua prepared the men to tramp off and claim their inheritance; but the old, the disabled, women, and children (besides whatever critters they had) stayed behind. Since the individual tribes did not know as yet what land they would receive they naturally did not build any permanent dwellings where they were. The men marched off to war, and the other folks stayed behind growing crops and tending livestock. As a plus, there is a water source located at the Jericho site.

Other archeological evidence indicates that the new inhabitants were wealthy and possessed of a new pottery and bronze weapons. No *single* people group prior to the Israelites lived in this land. The population was made up of a smattering of many different people groups. The evidence is that the newcomers were one people, and we are told that they brought with them the plunder of Egypt. Additionally, as were the customs of the time, it would be expected that the Israelites would have carried off anything of value from the conquered peoples east of the Jordan.

Shortly after the Israelites departed the hospitality of the Egyptians, that country was entirely swallowed up by the Hyksos and that without a battle. Put into perspective, it should come as no surprise. An estimated one-fourth of the population of Egypt had just left and Pharaoh's army destroyed as well (Ex. 14:21–28). It should also be remembered that the very area vacated by the Israelites was that corridor through pharaohs had marched to conquer and enemies had marched to conquer them. The Israelites had also helped themselves (as back pay) to the livestock and riches of the kingdom prior to departure. When the Hyksos took over, they went through the country destroying monuments, temples, and artworks while creating nothing themselves. They were a mongrel race that only took.

Additional Finds

There are those who claim that the Israelite kings David and his son Solomon are nothing other than myths of Israel's former greatness. They ask, "Why are there no contemporary statues, no records, no bas relief carvings to celebrate these two if they were so great?" Well, the reason for the absence of statues and carvings is very simple: it's called the Second Commandment: "You shall not make for yourself a carved image—any likeness of *anything* that is in heaven above or that is in the earth beneath, or that is in the water under the earth; you shall not bow down to them not serve them".

As far as records are concerned, they considered themselves as being in the process of writing *the* record, so scribbling on walls did not, I suppose, occur to them. However, in 1993, at the archeological site of Tel Dan in northern Israel, a ninth century BC stela was discovered that mentioned the House of David. The stela commemorates a military victory of an Aramean king over the "king of Israel" and the "king of the House of David." While the names of the kings were not preserved, it is believed that the victory described refers to a campaign between the king of Damascus (Hazael) and Jehoram of Israel and Ahaziah of Judah. What is perhaps just as important, the stela was erected one hundred years after David died, and by referring to his "house," he was still being recognized as the founder of the kingdom.

Let's talk about Assyria for a moment. Assyria was ancient long before David drew breath. It's original capital Asshur (the basis of the name Assyria) was established as early as 2,000 BC. It is said in Genesis 10 that Nimrod, great-grandson of Noah, went into Assyria (among other places) and established the cities of Nineveh and Calah. Ancient Assyria, like many other states, had its ups and downs during the approximately two thousand years of its existence. Fat times and empire sized one day and barely the size of New Hampshire the next. The thing about Assyria, however, is that their history is extraordinarily well documented and reveals itself to us today through royal archives, correspondence, *verifiable* lists of kings, statues, and other inscriptions of all kinds.

I bring all this up because the history of Israel/Judah regarding its interactions with Assyria as put forth in the Bible are mirrored by the archives and other materials recovered by archeologists in Assyria. When the palace of Ashurbanipal was excavated, we learned much from the over twenty-five thousand tablets recovered. Many of these records confirm the Bible's accuracy, and every single reference to a biblical king is confirmed in their records. Another object, currently in the British Museum, called the Black Obelisk of Shalmaneser depicts a Hebrew king (Jehu) or his lieutenant bowing before him in submission. One hundred and twenty-three years after the demise of Shalmaneser, his descendant Sennacherib laid siege to Jerusalem. He left a description of the event that also matches the biblical record, though, subject to human nature, he spins it his way. (He was unable to subdue Jerusalem, as foretold by the prophet Isaiah, but still needed to show his people at home that he was *the man*.) As transcribed, it says:

> As to Hezekiah, the Jew, he did not submit to my yoke. I laid siege to 46 of his strong cities, walled forts, and to the countless small villages in their vicinity. I drove out of them 200,150 people, young and old, male and female, horses, mules, donkeys, camels, big and small cattle beyond counting and considered [them] booty. Himself I made prisoner in Jerusalem, his royal residence, like a bird in a cage. Pritchard, ANET, as cited by Geisler, BECA, 52

Let's fast-forward to 626 BC, and we find Babylonia being stirred up. Nabopolassar, a chieftain, seized Babylon in that year. He established an alliance with the Medes and in 612 conquered Nineveh. Now, his son was the great Nebuchadnezzar who assumed the throne in 605 and twice besieged Jerusalem (598/97 and again in 587/86). The Bible tells us that during the second siege, the city was sacked, the temple destroyed and King Jehoiachin (along with his five sons) hauled off to Babylon. When the famous Hanging Gardens of Babylon were excavated, records were found that confirmed that the unfortunate Jehoiachin (and his boys) were, in fact, granted a monthly ration and a place to live and were treated properly if not kindly . . . just as it says in 2 Kings 25:27–30.

Genesis

All that we have discussed biblically this far has been post-Creation. There are many theologians who intentionally avoid the first eleven chapters of Genesis because they are living under the delusion of evolution or are otherwise intimidated by the overwhelming knowledge possessed by those great men and women in the hallowed halls of academia (including some seminaries, I might add). Of course, materialists merely howl in derision at the whole idea of supernatural creation and claim that any who entertain such thoughts are ignorant, backward, or worse. But if you have followed me thus far, we have shown that their explanations for why we are here, and where we came from, don't make sense; they just plain don't work. So what are we left with? The view left us is much more gratifying and encouraging that the thin gruel that we have been provided through the feeding tube of science or, more accurately, the materialist worldview.

You may think that here I am going to provide you with a hitherto unknown, secret revelation of *how* God created all things. The method is not any type of engineering or biological marvel; he spoke. Yep, that's it. He said what he wanted, and it appeared—instantly. He completed everything over six *twenty-four-hour days* and then took a break and admired his work. Could he have done it all at once? Of course, he could. He did this as a way of providing us with an example of how we should manage our time. He reiterated the scheme of resting on the seventh day in the Ten Commandments. Most people today recognize the need for balance among family, work, play, and rest in their lives to stay physically and mentally healthy and reasonably happy.

Now, before we move forward, let's look at a couple of questions that have been asked literally thousands of times. First, "Who created God?" This is a question suited to children but not to a thinking adult. A self-existing, eternal being has no creator. Now, of course, this does not prove the existence of such a being, but it does mean that we don't have to answer silly questions. If asked a question like that, the person that asks it simply does not understand what we have been talking about. This God is the First Unmoved Mover as so named by Aristotle. God is eternal, and he is the only thing that is eternal. Before he spoke into existence all things, there was *nothing* but God. (Actually, I find the concept of *nothing* extraordinarily difficult to comprehend.)

The next situation I want to address is that of folks saying, "Well, couldn't God have used evolution as his method of creation? Couldn't the six days actually mean billions of years? After all, the apostle Peter says one day is as a thousand years to the Lord." The short answer to this is no. Well, Dave, who are you to say so? Again, short answer: nobody. I give my response based on the text and something called the law of first mention. Well, what the heck is that? you ask. The law of first mention says (among other things) that the way in which a word is used first in a text is the expected use of the same word thereafter. Now the text specifically says, "So the evening and the morning were the first day." In addition, the word *day* is used in the same context of twenty-four hours more than 2,400 times throughout the Bible. (That squares it for me. If you want to debate it, go ahead.)

Okay, then, what about what Peter said in 2 Peter 3:8? Let's look at that verse: "But, beloved, do not forget this one thing, that with the Lord one day is as a thousand years, and a thousand years as one day". The context here is of Peter talking to his listeners about the *second coming* of Christ. He's reminding them to be patient and to go on with their lives. He is decidedly *not* talking about creation. One of the greatest dangers in reading scripture is the taking of a single verse or several verses, out of their intended context, and construing upon them a meaning that suits the individual.

Many people I talk to, following my response to the two previously mentioned situations, continue on with more questions. For example: "How could God have created the light on the first day but not the sun, moon, and stars until the fourth day?" Response: Initially, God himself provided the light. The sun, moon, etc., came later to make the planet suitable for our habitation as the text clearly indicates. After all, what use has God for seasons, days, years, day, and night? On and on it goes. It has been my experience that the vast majority of folks who ask me these types of questions have never read the Bible in full. Some people simply do not want to see. I remember that when I was in high school, I had absolutely no interest in chemistry. Guess what? I didn't learn very much. Most times these questions are bandied about by folk who have no sincere desire to find truth. But we are getting ahead of ourselves.

Having eliminated the theory of matter (evolution) from our list of viable options and having turned to the Bible, just what are we really hoping to find? We need, at this point, to remind ourselves what our goal is. We have reviewed the universe as matter and only matter. That is the universe of materialism, which is sometimes called naturalism, humanism, etc., and found their answers to our questions unacceptable. We are looking for truth upon which to base our view of the world.

The Ark

We have covered the geologic considerations surrounding a global flood previously. After the flood, all the earth was devoid of land-dwelling creatures, so there had to have been a way to repopulate. Since the only descriptive explanation we have is from the Bible, we are going to look at the feasibility of that narrative:

> And God said to Noah, "The end of all flesh has come before Me, for the earth is filled with violence through them; and behold, I will destroy them with the earth. Make yourself an ark of gopherwood; make rooms in the ark, and cover it inside and outside with pitch. And this is how you shall make it: The length of the ark shall be three hundred cubits, its width fifty cubits, and its height thirty cubits. You shall make a window for the ark, and you shall finish it to a cubit from above; and set the door of the ark in its side. You shall make it with lower, second, and third decks." (Gen. 6:13–16 NKJV)

Generally, Hebrew scholars believe that the cubit was not short of 18 inches. Using the minimum measure of 18 inches, the ark would have been 450 feet long, 75 feet wide, and 45 feet high. Prior to the start of metal shipbuilding in the nineteenth century, it was probably the largest seaworthy vessel ever built—maybe even the largest ever considered. Modern shipbuilders have said that it was so well proportioned that it would have been practically impossible to capsize. The total capacity of such an ark would be 1,518,000 cubic feet or the equal capacity of 569 railroad freight cars. Could this have been enough capacity for animals to repopulate the earth?

There have been credible books written to suggest that there was ample space for the family of Noah, animals, and forage. I don't know how many dogs or cats existed at that time, but one pair of each would have been sufficient. Larger animals could be included merely by the expediency of using babies. Nobody said that they had to be full grown specimens. It is estimated that between 16,000 and 35,000 animals would have been necessary. Let's go further to allow for error in the estimate and say 50,000 animals were present. Taking everything into consideration, the average size of the animals would be about that of a goat or sheep. Using our freight car as an example, 240 sheep can fit in 1 double-decked car. Only about 207 cars would be required to hold the animals, leaving 361 cars for the family of Noah, forage, baggage, food, etc. Of course, once the animals were aboard and the rains started, the work really began! It has been proposed that some or all the animals went into a state of hibernation, which would have saved a lot of feeding and shoveling time. Maybe but we'll probably never know. In any event, all the animals would have fit on one deck, leaving the family space and plenty of room for food.

Could Noah, his wife, three sons, and their wives have created the population we have on the planet today? (This same scenario will also describe how animals repopulated the planet.) Let's do some math using only girl babies as the example.

> Parameters: Each woman will have 3 daughters who reach maturity and has girls of her own at age 25.
> 3 25-year-old women leave the ark = 3
> 25 years hence 3 x 3 young women and girls of next generation
> 50 years later 3 x 3 x 3
> 100 years 3^5
> 200 years later 3^9 = 19,683
> 359 years later 3^{15} = 14,348,907
> 500 years later 3^{21} = 10,460,353,203 which equals more *girls* than the total number of *people* today.

This, of course, is not definitive by any means, but it does show that it *could* have happened. The math uses the same features as those of compounding interest. We know that there have been times of greater and lesser growth in population, but these numbers are still revealing.

Noah and his descendants were to go forth and fill the world. Well, people being what they are, his relatives decided to ignore the command and gather into a big herd somewhere around the future site of Babylon. God, being what *He* is, noticed what was going on and decided to intercede. You see, the people all spoke the same language (which makes perfect sense), had come upon an apparently attractive plain, and decided they would remain together and grow great. (Apparently, there were no Leif Erikson or Daniel Boone types in the bunch.) So here they decided to bake bricks and build a city. In this city, they would build a tower (infamous Babel) all the way into the heavens and would make a name for themselves all round about so that they would not be messed with. (I don't know who they were worried about.) In short order, God went down, told everyone to get out of the pool, and confused their speech, forcing them to move on and settle into pockets where they could understand one another.

What of this is substantiated by what we can find today archeologically speaking? Clearly, the events that occurred in the plain of Shinar were in the southern part of Mesopotamia. While the evidence postdates the event that we are speaking of, it can surely point us toward the event itself. There are around thirty structures in various parts of Mesopotamia called ziggurats (a ziggurat is a step-like structure built in stages). In the original Hebrew text in Genesis, the word *migdal* is used to refer to the tower built at Babel. Migdal is not a direct translation of ziggurat for the simple reason that ziggurats were not a part of Hebrew culture and they did not have a word that directly described a ziggurat. When one does not have a word for something, they typically use something descriptively similar. *Migdal* is usually used in a military sense to denote a "watchtower," and that seems to me to be quite close enough.

Many theories abound regarding why these ziggurats were built and what function they served, but the ones we observe today started to be built in the third century BC. Since they were built over previous locations, they probably appeared much earlier. All this to say that the original tower is thought to have been a precursor of the ziggurat. The biblical text mentioned that Babel was built specifically of burned bricks. It is interesting to note that kiln-fired brick is unknown in ancient Israel. There are none found in Palestine. In addition, the folks at Babel had asphalt (bitumen) as mortar where Hebrews used mud as mortar. This distinction is important as it points directly to the validity of the biblical text.

The Bible says: "Come, let us build ourselves a city, and a tower whose top is in the heavens; let us make a name for ourselves, lest we be scattered abroad over the face of the whole earth" (Gen. 11:4 NKJV). Knowing something of the culture helps us to fully understand what is happening here. In later Babylonian times, a city was typically comprised of the administrative and religious buildings often protected by a wall. Residents spread about the area outside the wall, demonstrating the first urge to urbanize. Further, at this time the general populace still lived a largely subsistence/agricultural existence. I mention this because the "scattering" would, of necessity, have happened anyway as evidenced by the story of Abraham and Lot having to part company because of lack of adequate grazing land:

> Lot also, who went with Abram, had flocks and herds and tents. Now the land was not able to support them, that they might dwell together, for their possessions were so great that they could not dwell together. And there was strife between the herdsmen of Abram's livestock and the herdsmen of Lot's livestock. The Canaanites and the Perizzites then dwelt in the land. So Abram said to Lot, "Please let there be no strife between you and me, and between my herdsmen and your herdsmen, for we are brethren. Is not the whole land before you? Please separate from me. If you take the left, then I will go to the right; or, if you go to the right, then I will go to the left." (Gen. 13:5–5 NKJV)

Perhaps the concern of God was meant more to hold off another disobedience. Maybe nothing was necessarily wrong in the building of cities and towers. After all, they certainly progressed with a vengeance over time. Might it not be more of what those cities and towers came to mean to the people themselves? Originally, the people strayed into idolatry by thinking that natural events were the representations of deity: thunder, lightning, earthquakes, etc. Later, the ziggurats came to represent stair-steps for the gods to descend to earth in order to get those things that they needed: food, water, rest, etc. In other words, gods were no longer seen as the force behind natural events but they had been reforged into the image of men. Perhaps this is what God sought to forestall in his destruction of the tower of Babel and the confusion of their language. Nothing is more sure of holding off urbanization than folks being unable to communicate, thereby eliminating their ability to efficiently work together. As the need for space increased, those who could understand one another would move off on their own.

Today, there are nearly seven thousand languages that we know of. These languages are grouped into language families. Vistawide World Languages and Cultures and Ethnologue (companies whose business it is to provide language statistics) agree that there are only ninety-four language families that have been identified to date. Maybe more will come. In any event, they fall well within the range of the people groups listed in Genesis 10. How could ninety-four languages have grown into the nearly seven thousand that we have today? Well, let's look at how language changes over time:

Our Father who art in heaven	Late Modern English (1700s >)
Our Father which art in heauen	Early Modern English (1500–1700)
Oure fader that art in heuenis	Middle English (1100–1500)
Faeder ure pu pee art on heofonum	Old English (c. AD 1000)

If someone came to me speaking Old English, it would be as foreign as Chinese. My cultural roots are in the southeastern portion of the United States. When I travel to Boston or New York, it is difficult for me to realize that we are speaking the same language (in some ways, I'm sure we're not).

The nearest I can make out by cross-referencing names with maps, the sons of Noah settled in the following areas:

> **Shem** ended up in Southwestern Turkey as well as between the Euphrates River and Lebanon, on the Tigris River, down around the Persian Gulf and Saudi Arabia.
>
> **Hams** folk trekked off to Southeastern Turkey, Palestine, Egypt, and Libya.
>
> **Japheth** found himself in Iran, Armenia, the island of Cyprus, the Greek isles, and Northern Turkey.

And so, people replenished and dispersed around the world. As subsistence was to remain a concern for centuries to come, the search for arable land and pasturage would carry groups farther afield.

Chapter Eight

NEW TESTAMENT RECORDS

Before we launch into the meat of our case, let's take a look at how some intelligent men in the past have chosen to look at "religion." We will not look at theologians because one can assume (many times in error, as it turns out) what position they hold. The learned gentlemen whose views we will briefly sketch here are positions that are still maintained today.

As a prelude to what follows, let's clear up what I conceive to be a misunderstanding. You remember Darwin's bulldog, Thomas Huxley? Well, he's the guy who came up with the term *agnostic*. While Huxley wished it to mean "no-knowledge" its general use is to mean "not enough knowledge." This is not going to work for us as it in no wise assists us in coming to any determination in selecting our worldview. Many people use it, however, simply to avoid making a decision on any number of topics, though, perhaps, most often on religion. In my opinion, if you claim agnosticism, you have already come to a decision by default.

David Hume (1711–1776) was a Scottish philosopher who also subscribed to the concept of agnosticism though he didn't call it by that name. He believed that no concept was amenable to rational discussion unless it could be verified by either definition, mathematics, or empirical methods. Mr. Hume did not believe that our sensual experiences were anything other than disconnected random events that provided no useful information in a search for truth. (Based on this, I suspect it was a good thing he never married. The poor woman would have had to talk to herself.)

It has been claimed by philosophers since Hume that he disregarded the law of causality. That is not strictly accurate. He said that you could not *rely* on it as a truth test. Basically, what he meant was that, taking the example of striking a cue ball with a cue, you couldn't know that it was the force of the cue striking the ball that caused the motion of the ball. How could we know that the motion did not derive from the action of the elbow or shoulder? In short, effects and causes come about sort of like water flowing from a faucet: in a steady stream and inextricably connected one to another and another and so on. In other words, my yodeling in a canyon today might have actually been the result of a Sumerian farmer hoeing barley in a field four thousand years ago. Any effect would have to be traced back through all previous causes to accurately ascertain the actual cause.

I don't know if Mr. Hume would agree with my assessment of his theory, and I can't ask him. He's dead. But it seems to me that if you follow his train of thought (or what his train of thought seems to me), this is the logical conclusion. In other words, all effects flow from a First Cause. Materialists who have adopted Mr. Hume's philosophy will howl at the conclusion I've drawn, but so be it: let them howl. What I conclude from this is that agnosticism is not an acceptable hiding place for anyone regarding the acceptance of one worldview or another. If agnosticism is to be allowed as a defense, how can the claimant profess any definite opinion on anything? One must have a foundation or be lost in the weeds.

There have been numerous philosophies put forth regarding ways in which we can *know* that there is a First Cause. (I'll call this cause God.) One of these is called rationalism and is based on the supposition that human beings have an innate knowledge of truth. Champions of rationalism were René Descartes (1596–1650), Baruch (Benedict) Spinoza (1632–1677), and Gottfried Leibniz (1646–1716).

The Rationalist Method

Descartes was led by the mathematical and geometrical progress of his day to use these tools as methods to articulate human reason. His starting point was doubt. From doubt, he deduces that the very act of doubting was proof that he was engaged in thinking, which, he concluded, confirmed his existence. His physical body, he said, could be doubted as it is only the extension of sensual experience. But since his mind believed in his physical existence and his mind was established and, furthermore, God would not deceive, then his physical existence was confirmed. This may seem silly on the surface, but remember, pantheists do not believe that they, or any of us, exist in reality. (I could have told Descartes he existed if he had asked me, but he didn't.)

Descartes did come up with a couple of useful things in this line such as his four rules for reasoning. They are (1) the rule of **certainty** would allow only undoubtedly distinct and clear propositions to be accepted as true, (2) **division** states that a problem must be reduced to its most basic form, (3) **order** provides progression from simplest to complex, and (4) **enumeration** requires constant verification of the argument to avoid error in reasoning. He is probably best remembered by the statement "I think, therefore, I am" being attributed to him.

While Spinoza was certainly a rationalist, he approached it from an entirely different direction. The basis of his argument was man possessing the ability to imagine the characteristics that would be possessed by a perfect being. In the presentation of his argument, he was primarily concerned with four ways in which humans err in their thinking process: (1) our being only capable of conceiving partial ideas, (2) the confusion inherent in our imaginations, (3) our abstract and nonspecific reasoning, and (4) our inability to develop a perfect concept of God.

Nevertheless, he proceeds to develop the concept of God as "necessary"—one whose existence and essence is contained within itself, having no contingencies. Further, he insists that some being must necessarily exist because *all* matters/events cannot be contingent; therefore, necessity and event/cause relationships demand that this being must also be infinite. The effects of this being are also, necessarily, infinite. This sounds pretty good on the surface, but in reality, his thinking ends up not in a God capable of personal relationship but a form of pantheism. God appears in a large variety of manifestations.

Gottfried Leibniz approached rationalism from yet another direction. He postulated a *sufficient* reason for God. He stated that it was known that all true propositions begin with the certainty that all effects have a cause. Further, he was a strict proponent of the law of non-contradiction. He also cited the principle of perfection as being a positive test for truth; since God is perfect, he has a moral obligation to create the best possible world. Leibniz used a combination of several elements in his ultimate argument for God. They include (1) the ontological (if one can conceive of a perfect being, he must exist), and (2) the universe is observed to be changing, and since it does not possess a sufficient reason on its own, there must be a First Cause (the cosmological argument).

These men, to be sure, wrote much more and have had much written about them, and there are certainly others who ascribe to the rationalist method, but what we have here noted is sufficient for our purpose. Rationalists have provided useful contributions to the search for truth. Among them are the confirmation of some basic laws of thought. For example, the law of non-contradiction cannot be disregarded without the absurd resulting. Also, without acknowledging the ability of the mind to grasp at least some things, any knowledge or truth would be impossible. Perhaps, most importantly, since anything real must be rational, it follows that reality can be known.

There are, however, drawbacks to an exclusively rationalist school of thought. For instance, just to be able to think something, of course, does not make it real. In addition, as regards the *necessary*, it cannot be proven rationally. Taking logic by itself, it is, at best, a test for what isn't, and logic alone leaves reality without any stuff. Logic and experience are necessary for a rounded view of the world.

There is another method that some use for searching out religious truth that is called fideism. Alvin Plantinga (1932–) defines fideism as an "exclusive or basic reliance upon faith alone, accompanied by a consequent disparagement of reason and utilized especially in the pursuit of philosophical or religious truth." Basically, fideism maintains that man possesses a natural and innate capacity to recognize the truth of God's existence in the same way that he accepts the truth of his perceptions (that is, a dog, a car, etc.). This is so *properly basic* that additional proofs are unnecessary. Along with this possessed sense, any objective argument may be overcome by an individual's subjective confrontation with Jesus. But since man is so thoroughly lost, he cannot recognize the truth unless he has had his eyes opened by the Holy Spirit. Man's reason, following this view, has a place but on a much lower rung of the ladder. Søren Kierkegaard (1813–1855) asked how it could be supposed that man could prove something through reason of which he could not even conceive.

Karl Barth (1886–1968) is an example of a more thorough-going fideist. He says that the only possible knowledge of God that we can have is that which is revealed in the Bible. God the Father is revealed through both the Son and the Holy Spirit. We are, in addition, so far fallen that we cannot even understand fully what he has revealed of himself so that we are only left with faith.

There are faults in the arguments of fideism as with rationalism. While the fideist may *know* that God exists, they fall prey to begging the question unless they can tell *how* they know he exists. In other words, they offer no test for the truth of their claims; and without a test, no truth claims can be made. How does one accept faith *in* God if one does not know that there *is* a God? Further, this method does not offer any truth claims but is reliant on faith alone so we are still left without knowing which faith-based claim is true.

Another methodology for establishing truth is called experientialism. Now, these folks say that only by extensive contemplating of your navel can you gain the truth of God. That, of course, is not really what they say, but that's how it sounds to me. Friedrich Schleiermacher (1768–1834) said that the experiencing of God would come through a disregard of traditional religious thought and focusing instead on "inward emotions and disposition." True religion is experiential; and all creeds, dogma, etc., were only the result of attempts to qualify the experiences themselves. The crux of Schleiermacher's doctrine is contained in his statements: "The whole religious life consists of two elements, that man surrender himself to the Universe and allow himself to be influenced by the side that is turned toward him in one part" and "That he transplant this contact with the one definite feeling within and take it up into the inner unity of his life and being. . . . The religious life is nothing else than the constant renewal of this proceeding." Wow.

Rudolf Otto (1869–1937) refined Schleiermacher by defining what is a true, justified religious experience. These steps or levels of experience he classifies as the *fearful mystery*. According to Herr Otto, true experience of the divine comes to us as follows: (1) the sense of religious dread or awe, (2) the sense of being overpowered, (3) an awareness of extreme energy or urgency that translates into sensational emotion (these three comprise his *tremendum*), (4) the person becomes blank from the awareness of the *Wholly Other*, and (5) while the Wholly Other repels, the beholder is drawn to the Holy through fascination with/of it (the five steps are full fearful mystery [*mysterium tremendum*]). Otto specifies the impossibility of describing God or the religious experience itself. Certainly, I would never be able to describe such an experience were I to have one. I'd have a stroke.

There are some serious drawbacks to relying on experiential methodology. To be sure, experience is certainly part of what gives meaning to faith. Without experience of God (and, to the Christian, the Son and Holy Spirit), we cannot adequately express the creeds on which we base our faith. However, even should you have an experience, it must still pass a test of truth. In other words, you can use an experience as the *basis* of your truth but not to *support* that truth. Furthermore, an event cannot define itself. For it to have meaning, it must be interpreted within the context in which the *individual* sees it. The experience itself is not group recognized, so even if the individual can articulate what happened, the same experience could be interpreted differently by others. What we are attempting to attain is truth, a world view that interprets all facts in the same way for everyone. Experiential methodology fails this test and proves itself to me to be meaningless, self-defeating, or, at best, begs itself.

The next method we will look at is evidentialism. Christians make claims on history that no other religions do. Specifically, a person named Jesus from Nazareth lived, was the incarnate Son of God, performed miracles to substantiate his identity, was crucified for his trouble, and rose from the dead three days later. Evidential methodology is based on the verification of historical claims. C.

H. Dodd (1884–1973) stated that the Christian faith was based on eyewitness accounts set forth in the New Testament. The authors of these books made no attempt to prove their testimony. They offered the information to be used for the edification of people who were contemporaries and those who would follow after them. In addition, New Testament scripture offers insights in fulfilling our mission as Christians and support for our faith in both hard and easy times. John W. Montgomery (1931–) possesses a legal background that he has used in putting forth his case for Christianity. Given the testimony of the apostles and the disciples, Montgomery points out that their statements as testimony would be considered "beyond a reasonable doubt" in a modern court.

Gary Habermas (1950–) is a distinguished scholar and professor who considers the evidence for Christ and his ministry to be firmly established via historical criticism. An historical event is comprised of not only the event itself but the meaning ascribed to it by its witnesses. Said another way, analogously, we should not use our point of reference today to describe and give meaning to the Battle of Gettysburg. It is far more instructive to listen to the descriptions and repercussions as delineated by those who lived through it. Christianity is not one event but a trail of continuous events from the birth of Christ through his resurrection and ascension. In addition to the events themselves, particular meaning is given them based on Old Testament prophecy and the lives that were changed because of encounters with both the living and risen Jesus.

Critics of this method raise a number of objections. For example, they cite the fact that no one living today observed any of the events recorded in scripture. They also point out a problem described as fragmentary accounts and that spin is imparted to testimony because of the use of *charged* language. It is also claimed that the witnesses at the time reported events based on what they were preconditioned to expect them to mean. Put another way, they claim that Christians selectively present their evidence because they are disposed to do so by being conditioned by their worldview. Last, but not least, the truth of miracles is unknowable. Of course, my reply to these objections in simple: they all apply equally to the "hard" sciences individually and materialism collectively, as well. Materialists also start with their own world view and the presuppositions adherent thereto.

However, there are valid objections to a strictly evidentiary methodology. Briefly they are (1) world views do contribute in evaluating the interpretation of a fact, (2) facts can be interpreted to have more than one meaning, and (3) reported miracles have to be discounted unless there is an independent method for verifying them. In short, while some objective knowledge can be obtained through evidence regarding the past, evidentialism left to itself cannot prove a Christian worldview as true.

Pragmatism is the concept that says truth is what works. As defined by *Merriam-Webster*, it is a "noun; a reasonable and logical way of doing things or of thinking about them that is based on dealing with specific situations instead of on ideas and theories." It was brought to prominence in the early years of the twentieth century. William James (1842–1910) stated his view of pragmatism this way: the statements "It is useful because it is true" and "It is true because it is useful" were equivalent with each other. Pragmatism, therefore, is not a sufficient test for truth simply because it can be construed as implying that truth is relative, which we already know is an absurd statement. Put another way, it could read, "It's true because it is useful to *me*," but it may not be useful to you. Also, it should be noted that there may be, indeed are, many things that work but that are decidedly not true.

Having reviewed these various methods for discovering truth and found them all wanting, we can easily become discouraged or fall back on the idea that truth cannot be known. But what all of them have lacked is an adequate test for checking their propositions along the way. If you have included any faulty ideas along the way, the entire opera falls apart. So what is an adequate test for truth? The best answer to that is one that I found offered by Norman Geisler (1932–) who is an American philosopher and the author of too many books to list. He's a smart guy.

All the previous methodologies were directed toward establishing one particular aspect of reality but were not helpful in understanding reality as a whole. As a starting point from which to understand all reality, there are three principles we will follow: (1) we will apply them to all of reality and not a specific part, (2) they will be self-evident, and (3) they will be undeniable and to attempt to deny them will result in admitting them to be undeniable.

In regard to point 1, we establish that something *is*. We do this be admitting the following points:

1) Being is. Being is undeniable since the one that denies exist (the principle of existence).
2) Being is being. A thing is what it is (the principle of identity).
3) Being is not nonbeing (the principle of noncontradiction).
4) Either something is or it isn't. There is no middle ground between existence and nonexistence (the principle of excluded middle ground).
5) Nonbeing cannot produce being. Ex nihilo, nihilo fit (the principle of causality).
6) Being causes being similar to itself. Like produces like (the principle of analogy).

Now, as we move forward, we must be very careful in what we consider to be an undeniable statement. Even truth statements (a pentagon has five sides) do not make the resulting event necessary. While a pentagon, by definition, does indeed have five sides, it does not, of necessity, exist. What we can deduce from this process, is that if we find undeniable statements within a worldview, it can establish that world view as true. From this, it follows that conflicting world views will be rendered false.

Another thing we must be careful to do is to make sure our interpretation of facts is consistent with the world view. Said another way, it would be absurd to insert God into a materialistic universe; there is no place for him there. Likewise, to reject miracles in a theistic universe would be equally absurd within the context of the system. Once an overarching framework is established for the interpretation of *all* the experiences and events in the world, it is the consistent fit within that framework that renders them truthful. Again, that system or framework must account for *all* the facts. Given our limited capacity, we cannot always and at all times correctly judge the level of consistency nor can we know all the facts. The use of probability (called system coherence) is used, therefore, to establish truth claims. What conclusions best fit the system in the most consistent way will be considered true. We will use this system to evaluate the truth of various theistic systems.

There are several versions of a universe that admits of a God, but the understanding of who and what he is varies a great deal (as you have probably come to expect). It is important that we review them, albeit briefly, because many of the famous have subscribed to one or the other, and it is very easy to become confused. In themselves, some of them seem to make sense, at least on the surface. After all, evolution seems to make some sense, at least on the surface. (Not really. Just kidding.)

As a result of the Renaissance and the ensuing Age of Enlightenment, a form of God worship called deism came into vogue. Basically, Newton's development of the physical universe as a purely mechanical system comprised of inviolate natural laws brought this about though certainly others contributed. Initially, it was comprised of five basic principles of religion developed by Herbert of Cherbury (1583–1648) that he assumed were common to all mankind. These were (1) there is one God, (2) God ought to be worshiped, (3) individual morality and reverence are the main components of the worship, (4) people are to be sorry for their sins and repent of them, and (5) the divine will reward people for their good and bad deeds. His principles included the idea of universal salvation (all dogs go to heaven) and that any revelation of God was both unnecessary and impossible.

There are some problems with this as in principle number 3. As noted earlier, just who determines what exactly morality and reverence consist of? Of course, the idea of universal salvation completely eliminates the notion of divine justice. The elimination of any value attached to the Bible leads one to ask just who this God that is worshiped is. Note the complete absence of mention accorded miracles; this is because it would effectually be a revelation of God and, based on the inviolate natural laws, this is impossible.

But from this beginning, deism degenerated into a barely concealed atheism. Those who followed Herbert were typified by Thomas Hobbes (1588–1679) who believed that religion was only a political force used to control the masses (shades of Karl Marx). John Toland (1670–1722) rejected anything supernatural, which makes one wonder why he claimed to be a deist at all (he was really a pantheist)! William Wollaston (1766–1828) merely said that Christianity had stolen the idea of miracles from pagan cultural myths.

Despite these various viewpoints, all deists, to be "officially" counted as such, subscribed to three propositions: (1) there is only one God who was the creator of the universe, (2) there are no supernatural events, and (3) God is unitary—period (they are having none of that Father, Son, and Holy Ghost business).

There are also those who profess to believe in a *finite God*. Plato believed that both matter and God are eternal and that there are things outside of God over which he exercises no control. God is basically a traffic cop who merely directs the coming and going of matter. God is pretty much limited to this sort of activity and is both limited in power and goodness. John Stuart Mill (1806–1873) sought to demonstrate this by pointing to the apparent prevalence of evil in the world. If God were omnipotent, he would, according to Mr. Mill, be able to control evil and suffering. According to Mill, matter and energy are both eternal, so God used material already available to form the universe. After God's job of creating was completed, the inviolate natural laws took over and God was out of a job.

In 1981, an interesting book came out written by one Rabbi Harold Kushner (1935–) called *When Bad Things Happen to Good People*. Kushner too takes the line that God is limited in both power and perfection. While God is indeed the creator of the universe, he is limited not only by natural laws but also by human nature and free will. God does not cause catastrophes to come among us, but he is also incapable of stopping them. He does provide us with strength to overcome circumstances, but he cannot control them. Kushner does credit God with providing moral laws and states that God does expect our obedience, but this obedience is not the highest form of morality. To him, the highest moral order comes when people who have the worldly experience change the rules to better serve society. As time passes, people will come along with better rules and will thereby continue to improve the world. (Yikes! That sounds familiar.)

Basically, those who believe in a finite God conform to the following points:

1) God is finite in power, compassion, or both.
2) Evil is real and cannot be controlled/defeated by God.
3) God did not create the universe but formed it from preexisting materials.

4) Natural laws rule.
5) Miracles do not happen because they violate the natural laws.
6) God used an evolutionary process for human development.
7) Ethics/morality are relative.

This concept of God has some serious problems. For example, if God is finite, that means he is a contingent being dependent on something/someone else and therefore not God at all. Someone greater than he had to have created him and that greater something/someone would be god. Also, if God is not all-good, then how do we measure just how good he is? If there is a greater goodness by which to compare him, then that greater goodness would be God. It also seems clear to me that it is entirely absurd even recognizing the existence of this god. After all, he is incapable of helping us, he couldn't even create us, and it is obvious that, since he is unable to do anything about evil, that it will continue to plague us. Heck, this god is unemployed for crying out loud!

While most of us think of polytheism as having to do with Greek and Roman mythology, there are some modern versions currently held. The Greek and Roman hierarchies are actually referred to as henotheism because they have a head god—sort of a lead dog to keep the others in step: Zeus in Greece and Jupiter in Rome. Polytheism as practiced today came about largely because of the "death of God" in the 1960s and 1970s. With this untimely death of God, some sought to fill the vacuum with an odd assortment of gods and goddesses. This assortment includes Hinduism, Mormonism, Wicca (modern witchcraft), and Buddhism.

The case of Mormonism is very interesting. Joseph Smith (1805–1844), founder of this cult, claimed to be *the* new-day saint and reported that he had received numerous messages from God conveyed to him via an angel named Maroni and some others in the know. These revelations ultimately found themselves published in the Book of Mormon. Mr. Smith revealed that God is the ruler of this world and started out as a man. The god of our world has a wife, the Heavenly Mother, who assisted him in begetting all that has been begotten. All souls have eternally existed, and they are given bodies when a newborn child requires a one. People spend their lives in an attempt to attain the Godhead just as God and the

Heavenly Mother did before them. (It is to be assumed that they are awarded another planet to rule over.) Godhood can be attained by possessing a plurality of wives and procreating in heaven. Presumably good works also aids in the effort. There are currently three hundred million such gods and counting.

There is kind of an interesting sidebar to modern polytheism. George Lucas, who follows the Zen Buddhism discipline, has ingeniously inserted these beliefs in the *Star Wars* movies under the guise of "the force." The force can be both good and bad, and the battle to attract adherents is waged for eternity by both sides. There are no ultimate winners, only losers, because there is no supreme being to sort them all out. In this view, mankind, through the goodness of their collective hearts, must merely strive to do the best they can, hoping that the good force will ultimately win.

There are basically three features to modern polytheism, and they are the following:

1) By definition, the notion of one god is rejected.
2) Relativity plays a major role in morality.
3) Irrationalism rules the day.

It is only fair to point out that this is a fairly small "religious" movement, at least in the West.

So that leaves us with theism. What does *theism* mean? Once again, we'll count on *Merriam-Webster* to tell us. It says: "The belief that God exists or that many gods exist: *specifically*: belief in the existence of one God viewed as the creative source of the human race and the world who transcends yet is imminent in the world." I'm not sure why they include in the first instance "or that many gods exist" simply because we've already discovered that there is a specific word to reference this concept, polytheism. We will go with their *specific* definition.

The principles of reality that we have discussed earlier draw several conclusions that lead logically to the end that this God exists. (Remember old Aristotle? This was largely the argument he stated so long ago.) These conclusions can be stated this way:

1. It is undeniable that some things exist.

2. As a finite being, my nonexistence is possible.
3. If it is possible not to exist, the potential to exist must be actuated by something/someone else.
4. An infinite regress of causes is impossible.
5. There must be an uncaused First Cause in order for things to exist.
6. The First Cause possesses those characteristics represented by the God of theism.
7. The God of theism exists.
8. The God of theism is the same as the God of the Bible; therefore, the God of the Bible exists.

Naturally, atheists/materialists have a number of criticisms to offer regarding these principles, and just for the sake of fairness, we'll mention a couple of them here:

1. Some simply attempt to sidestep the principle of causality by postulating an infinite number of causes. This, however, would require all effects to have been self-caused (which is impossible) or uncaused (which is absurd). The fact that some things *are* necessitates something to make them be.
2. The principle of causality is justified as an argument because that which is potential cannot actualize itself. While it is true that it could have been that I would not exist, I do, so it follows that some existing effect brought me into being.
3. Atheists argue that all the world did not require a First Cause. This would make sense if all existence were not made of the same stuff. Every single thing in existence requires a cause just as much as any other single thing. They further want to claim that the whole is more than the sum of its parts, but this actually becomes an argument *for* theism rather than a refutation. It "the universe" is thought of as the total of all contingent, finite beings then there must be a cause to have brought them into being. Should the universe be considered that which is eternal, necessary, and more than the sum of all the effects, then that is ultimately what the theist means by God.

4. Some atheists want to claim that the terms *necessary being* and *uncaused cause* have no meaning. They would say that necessity has nothing to do with existence. While God may not be a *logically* necessary being, the statement "God exists" can be a logically necessary *statement*. In other words, logically necessary statements about existence can be allowed. Further, while a necessary being may not be *logically* necessary, it does not exclude God as an *actually* necessary being.

What does the Bible tell us about the world we live in, the reasons for it being so awfully messed up, and what, if anything, we can do to make it right? If we can't make it right, can we at least find some peace and happiness while we are here? We'll look at that next.

We should be convinced at this point of the reliability of the Old Testament text as well as its historical reliability. The New Testament is even more extensive in the amount of manuscripts available that testify to the accuracy of the book we have today. Currently, there are over 25,000 whole or partial, ancient New Testament copies. Compared to other ancient writings, this represents a ridiculous amount of material and, perhaps more importantly, the short interval of time between the copy and the writing of the original.

We can get an idea of just how strong this manuscript evidence is by comparing it to other ancient writings. For example, Julius Caesar's *Gallic War* (composed between 58 and 50 B.C.) has only 9 or 10 good copies in existence and they are from 900 years after his rule. Only 20 manuscripts of Livy's history of Rome survive of his 142 books and the oldest is from sometime during the fourth century. The Histories of Tacitus (c. A.D. 100) come to us through only 4 ½ books and his Annuals consist of only 10 of his 16 volumes and his two greatest depend entirely on two manuscripts dating from the ninth and eleventh centuries.

The History of Thucydides comes from manuscripts belonging to c. A.D. 900; the History of Herodotus from the same time. 1,300 years later than the originals. No scholar today, however, would consider the publishing of these writers as anything less than authentic. In short, the New Testament is far and away the best validated ancient writing by the very number of documents available to us and the short time between the originals and copies.

Ancient authors and their works compared to New Testament manuscript availability.

Author	Book	Date Written	Earliest Copies	Time Gap	No. of Copies
Homer	*Iliad*	800 BC	c. 400 BC	C. 400 yrs	643
Herodotus	*History*	480–425 BC	c. AD 900	c.1,350 yrs	8
Thucydides	*History*	460–400 BC	c. AD 900	c. 1,300 yrs	8
Plato		400 BC	c. AD 900	c. 1,300 yrs	7
Demosthenes		300 BC	c. AD 1100	c. 1,400 yrs	200
Caesar	*Gallic Wars*	100–44 BC	c. AD 900	c. 1,000 yrs	10
Livy	*Hist. Rome*	59 BC–AD 17	mostly 10th century	c. 1,000 yrs	19
Tacitus	*Annals*	AD 100	c. AD 1100	c. 1,000 yrs	20
Pliny Secundus	*Nat. History*	AD 61–113	c. AD 850	c. 750 yrs	7
New Testament		AD 50–100	c. 114 (fragment) c. 200 (books) c. 250 (most of NT) c. 325 (complete NT)	50 yrs 100 yrs 150 yrs 225 yrs	5,366

(chart from Ravi Zacharias, *Can Man Live without God*, 162)

Another indication of the existence of all twenty-seven New Testament books in complete form very early on is in the number of direct quotations by the early church fathers. There are certainly times where the references are in paraphrase form, but the quotations are so numerous and were so commonly used that if no books were in existence of the New Testament, the complete books themselves could be reproduced from the quotations alone.

Below is a chart of the early patristic (writings of the early church fathers; derived from the Latin and Greek for *father* [*pater*]) quotations from the New Testament:

Writer	Gospels	Acts	Pauline Epistles	General Epistles	Revelation	Totals
Justin Martyr	268	10	43	6	3 (266 allusions)	330
Irenaeus	1,038	194	499	23	65	1,819
Clement of Alexandria	1,107	44	1,127	207	11	2,406
Origen	9,231	349	7,778	399	165	17,992
Tertullian	3,822	502	2,609	120	205	7,258
Hippolytus	734	42	387	27	188	1,378
Eusebius	3,258	211	1,592	88	27	5,176
Grand Totals	**19,368**	**1,352**	**14,035**	**870**	**664**	**36,289**

Of course, there are still those who want to jump right away to supposed contradictions, errors, and so forth they claim are in the text itself. Generally, I refer back to my earlier comment that most often these criticisms are offered by people who are not familiar with the Bible as a whole and/or simply want to argue based on a contrary worldview.

So far, we haven't discussed what the New Testament actually tells us, and many people want to advance the idea that self-interested people just made the whole "story" up. Before we get into stating the case for a God-centered world view in depth, let me just take a moment to answer people who believe so silly a statement. We are told that when Jesus was arrested, his friends ran away and left him alone with the mob who came to take him away. Why did they run away? My guess is that they ran because they didn't want to be arrested. But the real question is, what made them come back? What made them all decide to give up the faith and homes that they had known all their lives to suffer poverty, homelessness, and, ultimately, torture and death (aside from John who was banished to the island of Patmos)? All this for a lie? Human nature would say that at least one of them, while alone in prison or facing the ultimate sacrifice, would have recanted their story to preserve their life; yet none of them did.

So where are we at this point? We have discussed materialism/evolution and found that it provides answers but not good ones; they are simply made up. In addition, we know what the meaning is behind this worldview. It leaves us with hopelessness, gloom, and despair simply because we are only a conglomeration of goo, accidentally adhering together to form *us*, providing the individual with no meaning and no future. Fortunately, I came to see the lie on which this world view is based, and as we move forward, my hope is that you have too.

Let us here pose to ourselves a question that is of supreme importance (or at least should be) to all of us in making decisions that directly impact our lives. That question is, "What is truth?" It has become very popular today to say that "truth is relative." But what does that mean? Relative to what? The very statement is absurd. "Truth is relative." Is the statement true or is it relative? Most people expect truth from their spouse, kids, doctor, lawyer, etc., and they will not for a minute accept any "relativity" from them. But when it comes to establishing the truth of a world view, many recoil and take the position that one view is as good as another. What works for you may not work for someone else. But again I say *everything* cannot be true, which is exactly the inference of this proposition.

Maybe the real problem is when we seek to resolve issues or truth claims about religion. Can we, in fact, discover any truth about such a topic? If there is only opinion, then we are right back where we started. Should there turn out to be truth claims that can be made about one religion or another, it necessarily follows that all other religions are false. At this point, the cries are resounding: "Who are you to say any religion is false?" "That's incredibly intolerant and pretentious of you!" Response: what does pretention or tolerance have to do with *truth*?

We have already decided that materialism is a dead end as well as monism/pantheism. The only option left open to us is that which stems from accepting a First Cause. Can we know anything about this being? If so, then how? Because this is so critical to our journey, we will have to adopt a very careful study to ensure that our findings are sound. We have had something to say of Old and New Testament historical reliability, but we have not at this point put them together. The glue that holds them both together and provides significance for us is through the illumination and revelation of Jesus Christ. It is all about him from Genesis 1 through Revelation 22. It is all about him. We are the unbelievable beneficiaries of His existence and work which the Comforter's opening of our eyes to the truth and the Father's mercy will help us to see.

Chapter Nine

STATING THE CASE

The Bible is the story of Christianity. The Old Testament is the Jewish Bible, but it is incomplete. The New Testament completes the promises of God made in the Old, or better said it is both an end and a new beginning. First, it reveals how we came to be and, throughout its contents, reveals some wonderful traits of the Being who created us and everything else. It is also descriptive of our failings as his creatures and provides many examples of our collective natures. In one aspect of character or another, we can identify with many of the historical people named. Maybe we possess *all* of their characteristics both good and bad. We do, I think.

The most important feature, however, is that it is true. And I don't mean *relatively* true but true for all people, at all times, and in all places. It is a true revelation of reality. It does you no good to cherry-pick the *truisms* that suit you. It is not an allegory. The people named really lived. The events depicted really happened. We've already reviewed other "belief" systems and found them not only wanting but also wrong—impossible, in fact. So let's look and see just how the Bible resolves the questions we want answered in order to firmly establish our world view.

We established earlier that God created us. Genesis does not tell us why, but we will be able to figure that out as we go along. He created it all out of nothing. He had not stockpiled lumber or fill dirt and possessed no nursery to retrieve the plants from. He is our First Cause. He is eternal because he was here when there was nothing else. Further, he is omnipotent and omnipresent. The Bible also tells us that he's caring and loving. In fact, we'd know nothing of love without him. How do we know this? Because all his created life-forms were placed in an environment free of predation, provided with all they'd need to prosper, and he personally cared for them. This is the very picture of a loving Father.

He made man and woman. The first were named Adam and Eve. It is significant to our special relationship with him that he did not speak them into existence but fashioned each with his own hands. We are also made in his image. We were given God's communicable attributes: reason, intellect, will, and emotion. At the time, we were also in his image morally in that we were good (very good, in his words) and sinless. But do not be puffed up with our being in his image. In the original Hebrew, the word used for *image* is *tselem*, which indicates a shadow or just an outline or representation of the original. We are as much like God, in other words, as our shadow is like us.

But he did more than this for Adam and Eve. He gave them dominion over the earth and its creatures. In addition, they could roam, play with the animals, frolic, and procreate. They were free. They were naked (the climate, I'm sure, was conducive) and unashamed. Their creator was their friend. They could do anything that they desired—except eat of the tree of knowledge of good and evil. He loved them enough to set some boundaries but not to prohibit them from free decisions. *The* tree was not wrapped in barbed wire and guarded by angels with automatic weapons. They were given the freedom to choose their own path. (Now, I can hear the questions popping but be patient. We'll have Q&A in a while.) They were also given God's model for the family: one man joined to one woman.

There is something else that needs to be brought to your attention. God created the universe and all that is in it. That means everything belongs to him. While he gave us a comfortable place to live, some fun stuff to do, he did not primarily make all this for us. He made it to manifest his glory. This is his kingdom. That may rankle some folks, but it is what it is. We are his subjects, if you will (even if you don't will). Now, the truly great thing is that we are not ruled over by some clown of an earthly king (or even a good one), or a group of pompous, narcissistic politicians but are looked after by a being that possesses all the attributes we mentioned earlier. He is, in fact, more like a father who is also a king, but we should never forget who is in charge. That's when things go sideways.

But there is more. God differentiated us from the other creatures. He made us special. We possess an inherent value as being made in the image of God and belonging to him. Human beings are the most special things next to God. But we are not minigods and goddesses. We were created, and created beings we will remain. We are inclined to make ourselves the center of the universe, and this is another mark of ingratitude to our Maker. This turns life into being all about you or me or whoever, not about God. Remember, it's about him, not us; we are his special beneficiaries, true, and we are special and have a value attached to us that is beyond comprehending as we will soon see—but it's still about him.

We also, because of our specialness, have duties to one another besides those due God. We are *all* special and God's creatures. This should be kept in mind when we interact with one another. Needless to say, people in general do *not* keep this in mind, but we *should*. And here I'm going to say something that I know some will not like. *If* we recognize ourselves as belonging to God, how then do we claim that we also belong to ourselves? This is an obvious contradiction. *If* you belong to God, your body is not yours to do with as you wish. We've already seen that you *can* behave as you want, but your actions have consequences. You know where I'm going with this. All human life is special and belongs to God who made it possible, yours, and the unborn child you are carrying; and if you claim God, you have no right to dispose of that special life that you have been blessed to bear. Personally, I hold that you have no right even if you do not profess God. I say this because I know folks who claim Christ who entertain the thought that a "woman has a right to choose." I personally do not see how you can belong to God and still claim any such thing as a "right" that belongs to you.

Lest anyone think that I am picking on the female of our species, I believe this also applies to any harmful activities that we choose to participate in. Overindulgence in alcohol, drug use, smoking, playing with rattlesnakes (I don't know where that one came from), etc., are all examples of playing with someone else's money. In short, we need to be consistent with that which we profess to believe.

Sometime later, we don't know how long, Adam and Eve fell prey to the greatest lie of all time—"You will not surely die"—and they ate of the infamous tree. The damage was done; they had broken faith with their Creator. They had decided that the very one who had made them didn't need to be obeyed, that he didn't have their best interests at heart, and that they were capable of deciding such things on their own. In an instant the relationship was broken, sin and self-will entered the world, and man entered a most lamentable state.

Here, we must mention one of the other attributes of God. He is holy. Holy is perfect and all-good, and sin cannot be allowed a pass. Sin, simply stated, is disobedience to God because a holy God, to be such, must also be just. Justice to his holiness must be served. (More on that later.) At any rate, Adam and his lady were hustled out of the resort, and the world as we know it began. The balance of the Bible is about how this good Father, despite our continued self-centeredness and with no credit do us (as we are reminded by the text), put into motion his predetermined plan to reclaim us.

You see, it isn't some abstract notion of the "world" being broken but the hard fact that we—all of us—are broken. We are special, certainly, but also doers of evil. Through our ancestors and continued by us, our world has become what it is, and, sorry, none of us are innocent of participation in this debacle. Today, we are pleased to speak from soapboxes about ignorance, poverty, lack of opportunity, crime, or environment as the causes of so much woe. These are merely the manifestations of the root cause. All these conditions are the result of evil either from direct participation in or being apathetic toward the perpetration of it upon others.

Whoa, there, Dave. I wasn't there. My wife and I don't even like apples. Well, I beg to differ. Everyone was there. I'm sure you enjoy your life of freedom. Even though you have probably made some bad choices (some worse than others) along the way, you would hardly be willing to trade freedom for the existence of a mindless robot. So one of our first choices was to rebel against our Maker, and don't pretend that we aren't still at it today. All of us. Remember, our benchmark is

Holy God. To him, the thought of murder is the same as the deed. To a Holy God, to dislike our brethren is the same as hating him. Our hearts are wrong. Our minds are wrong.

Look at it this way. Where do you think our genes came from? And our genes have now been watered down by thousands of years of human contrariness. Even evolutionists believe that we all came from a common ancestor. (For me, it's much clearer to visualize myself as descended from a man and a woman than from a squid.) In fact, even the great David Hume argued for it, albeit unintentionally. Remember, he basically said that to trace the actual cause for any event, we would have to go all the way back to the beginning because all events are merely individual experiences mixed up together and undefined. Besides, who really thinks that put in the same position, they would have behaved any differently. I mean, get over yourself.

Let me ask you a question. Do you believe that man has improved over time? Have you noted any change for the good in his nature? His handling of the world? Obviously, I'm not talking about the improvement in his toys (that seem to distract so many of us) or all the wonderful conveniences he has invented to make our lives more comfortable. I recognize the enormous contribution to society that the inventor of the electric toothbrush has made. And certainly, I would be remiss if I left out the automatic ice dispenser on the refrigerator door. And don't forget the longer-lasting lightbulb. No, I'm not speaking of these worthy contrivances. I'm talking about *man* himself. It seems to me as though man has, over the centuries, merely ratcheted up his ability to create misery. To be sure, there are exceptions to this. But the very way we speak of them as exceptions proves my point. They are unusual because they stand juxtaposed to the vast majority of the reported events we have thrust upon us and that we at times witness ourselves.

No, sir, the human condition has continued to deteriorate the further we have been separated by time and inclination from the garden. Every conceivable perversion and depravation known to man is not only conducted in the dark places but openly for all to see. In fact, try to count the number of things that we do not even consider perverse any longer—at least as a society. Has all this happened because we've *progressed*? Many of you might ask, "Dave, if God is so loving, good, and powerful, why has He allowed all this to go on?" (I believe that this is another instance of other-than-me responsibility.) Maybe you're thinking that God is not powerful enough to stop evil or that he lacks sufficient goodness. These are fair considerations, and I think we should take a few moments to look at them even though I had earlier asked you to hold your questions.

Power does not really have anything to do with it. When we say that God is omnipotent, we are saying that he possesses the power to enable him to do whatever can be done with power. In other words, God is not capable of doing irrational things. He is perfect in all his attributes, and one of his attributes, one that he passed on to us, is rationality. So the old questions of "Can God make a round square?" or "Can God make something so heavy even he cannot lift it?" do not have any meaning because they are absurd. God cannot do *everything* because many such propositions are simply silly and God doesn't go in for nonsense statements. God cannot do the irrational. And this should give us *greater* confidence in him than otherwise. I mean, how comfortable would you be with an *irrational* God? Someone who could willy-nilly change his mind and do the opposite of what he had promised? This is *not* the one we want to put our confidence in. So it isn't power.

We then ask whether an all-good God is *always* able to prevent evil at least as far as he is able. On the surface, this seems to be a nonsense statement itself as we immediately answer, "Yes, of course." But not so fast, my friends. Can you think of any situation where a good person would allow a small bad if it turned out that the end result was a larger good? Have you, as parents, ever allowed your child to endure something hurtful knowing that by doing so he or she would be better for having gone through that hurtful experience? Viewed in this light, can we not say with reason that God will prevent evil unless he knows that a greater good will be the fruit of it? Indeed, might it not be that he will allow a lesser evil to prevent a greater? If these statements have the ring of truth, then God's power and goodness are not really in question, are they?

So we are cut off from relationship with our Creator and left to our own devices. Just what did this mean?

> To the woman He said: "I will greatly multiply you sorrow and your conception; In pain you shall bring forth

children; your desire shall be for your husband, and he shall rule over you." Then to Adam He said, "Because you have heeded the voice of you wife, and have eaten from the tree of which I commanded you, saying, 'You shall not eat of it': "Cursed is the ground for your sake; in toil you shall eat of it all the days of your life. Both thorns and thistles it shall bring forth for you, and you shall eat the herb of the field. In the sweat of your face you shall eat bread till you return to the ground, for out of it you were taken; for dust you are and to dust you shall return." Gen. 3:16–19 NKJV

There is a noticeable difference between these words and the situation in the garden that God was pleased to put them in originally. While residing there, all was comfort and ample provision. Death, pain, and anxiety were unknown. Now, they get the whole truckload. God kills animals (the first death in the world) to clothe their previously unnoticed nakedness. Shame and lustfulness henceforth will be commonplace. Instead of plucking fruit from the trees, man must now work for his food; he will be subject to bad harvests and famine. God holds Adam responsible for listening to Eve. This is not a swipe at womanhood—after all, God made Eve as well—but rather an indictment of Adam for listening to *anybody* other than the one who made him. And, of course, Adam is quick to blame God in response for having given her to him. Discord, disunion, and separation. We have either been trying to sneak back into, or utterly destroy any memory of, the garden ever since.

And as far as that oily serpent is concerned:

> So the Lord God said to the serpent: "Because you have done this, you are cursed more than the cattle, and more than every beast of the field; on your belly you shall go, and you shall eat dust all the days of your life. And I will put enmity between you and the woman, and between your seed and her Seed; He shall bruise your head, and you shall bruise His heel." Gen 3:14–15

This turn of events did not cause God to have an "aw, shucks" moment. Already, God is giving us a view of his plan for reconciliation. Notice that Eve's seed is capitalized. Only proper names and references to God are capitalized. In addition, we are told that "He shall bruise your head." In other words, a future manifestation of God will strike a mortal blow against Satan while Satan will only cause him a temporary injury. This tells us even more because God as spirit cannot be injured, so the future manifestation must be more than spirit. He must be an incarnation. But I am getting ahead of myself.

In the Old Testament, we're told just how well folks did off on their own. So well that God needed to scrub the whole planet clean with the exception of Noah and his crew. Eventually, a man named Abram came along, and God was ready to begin the process of rescuing us from ourselves. From the seed of Abram (Abraham) came the Hebrews who were the folks chosen by God for a new beginning. The Patriarchs (Abraham's sons) gave birth to the twelve tribes; and they, after some hundreds of years, found themselves the slaves of Pharaoh. A leader was raised by God, Moses, who conducted them out of bondage. While wandering in the desert, they stopped to visit Mount Sinai (all the tourists go there) and thereupon Moses received the Law—the Ten Commandments. It is worthwhile to check them out:

1. You shall have no other gods before me.
2. You shall not make for yourself a carved image.
3. You shall not take the name of the Lord your God in vain.
4. Remember the Sabbath day, to keep it holy.
5. Honor your father and mother that your days may be long in the land.
6. You shall not murder.
7. You shall not commit adultery.
8. You shall not steal.
9. You shall not bear false witness against your neighbors.
10. You shall not covet. Gen 20:3–17

If you look them up, it will be apparent that I have paraphrased radically. There is much additional information provided though the above is sufficient for our purpose. God provided Moses with additional laws relating to servants, violence, animal control, property, justice, and morality. There were also various ceremonial and religious observances put forth, dress for the priests, the recipe for anointing oil and incense and such. While Moses was still on the mountain, in the very act of receiving these laws, the people were misbehaving on the plain down below. And so it has gone.

Over the centuries, most Hebrews forgot their God or at best offered only external obedience to rules and regulations. They eventually asked for a king to rule over them, and God gave them one. David became a great king; and Solomon was wise, rich, and famous; but they were not able to restore the broken relationship. They were men after all and prone to misbehaving. The glory days of their kingdom didn't last long. Over the centuries, the experience of kingship was good and bad——mostly bad. They possessed prophets who told them the destructiveness of their behavior, but they were generally ignored or even killed though some decided that their writings needed to be preserved. (How ironic is that?) It finally reached the point where their kingdom ceased to exist, and the people that survived were scattered.

But the Hebrews had been told to look for a Messiah who would come and make things right. Over time, with all the bad kings and oppressions perpetrated upon them by other nations, this Messiah came to mean more and more an earthly king who would throw off their shackles and reestablish them as a power to be reckoned with—never to me messed with again. But they had missed the message of the Ten Commandments, the prophets, and all those bad kings. The Jews were looking for someone to set them right with the world when what they needed was someone who could make things right with God and restore humankind to a right relationship with Him. Without that, the world couldn't be made right. The realization of whose kingdom it is had long since been swallowed up in legalism and false teaching.

After the prophet Malachi, the written record went blank and remained so for four hundred years.

A New Beginning

> Now in the sixth month the angel Gabriel was sent by God to a city of Galilee named Nazareth, to a virgin betrothed to a man named Joseph, of the house of David. The virgin's name was Mary. And having come in, the angel said to her, "Rejoice, highly favored one, the Lord is with you; blessed

> are you among women!" But when she saw him, she was troubled at his saying, and considered what manner of greeting this was. Then the angel said to her, "Do not be afraid, Mary, for you have found favor with God. And behold, you will conceive in your womb and bring forth a Son, and shall call His name Jesus. He will be great, and will be called the Son of the Highest; and the Lord God will give Him the throne of His father David. And He will reign over the house of Jacob forever, and of His kingdom there will be no end." (Lk. 1:26–33)

By the time Jesus arrived on the scene, Jewish faith had become mostly ritual, at least to all but the common man. The Great Sanhedrin consisted of the Nasi, who was the president, and a varied number of members depending on the reason they were meeting. The Sanhedrin adjudicated many matters civil and all matters religious. Members were largely selected from the parties of the Pharisees and Sadducees. The Pharisees were primarily concerned with maintaining control through keeping their religious rituals and rules pure while the Sadducees were more politically motivated. Neither of them, generally, were concerned with the souls of the people or in assisting the poor and hurt. The consensus was that they were getting what they deserved. Needless to say, most, if not all, of the ruling class were well off by the standards of the day. In fact, the less fortunate had been trained to view themselves as sinful because if they were possessed of any righteousness, they too would be well-off. In short, it was believed that God blessed the righteous with material well-being; therefore, if you had a lot of stuff, you were "good" and if you didn't have it you were "bad."

It is not to the purpose to review the birth of Jesus. That he did live is a fact as far as any historian worthy of the title will testify based on the evidence. What we need to carefully consider is *who* he was and *what* he did. He was born into this world through Mary's pain, labor, and blood in the same fashion as all men since Adam. In childhood, I imagine he went through many of the growing pains familiar to all of us. He bled when he was cut, needed oxygen to survive, and cried when he hurt himself. He was sad when he wasn't included in games. In physical existence, he was one of us. He was a man.

But he was more. During his life, he told crowds that he was deserving of the same honor as God. He said that he had existed with the Father before he was born. He possessed living water. He called on people to trust in him and never die. In the end, he said the judgment of all mankind has been left to him. These are not things that an ordinary Jew would even think let alone speak out loud. This was blasphemy. Lest we find ourselves believing he was merely extraordinarily egocentric, consider that he backed up what he said. He fed the multitudes, healed lepers, gave sight to the blind, and raised people from the dead.

What we are saying is that not only was he a man but God as well. How can this be? To get a sense of what we're talking about, let's take a look at the gospel according to John:

> In the beginning was the Word, and the Word was with God, and the Word was God. He was in the beginning with God. All things were made through Him, and without Him nothing was made that was made. Jn. 1:1–3

Let's compare this to what we're told in Genesis 1:1.

> In the beginning God created the heavens and the earth.

The conclusion that we come to is that God and Jesus are one and the same. Furthermore, we are told in Matthew 28:19:

> Go therefore and make disciples of all the nations, baptizing them in the name of the Father and of the Son and of the Holy Spirit.

This verse adds a third personality of God in the form of the Holy Spirit. This introduces God as the Trinity, a name never used in the Bible but clearly inferred. How does one reconcile this? The early church attempted to articulate a definition, but since it is outside human understanding, we will not attempt a theological understanding here. Basically, God is three separate personalities in one person. They are capable of relationship with one another but still one. I can come up with one example to explain it, but it is obviously a gross oversimplification. I'll give it to you anyway.

I am one man named David Tozar; but my wife, children, and grandson all know me differently. To them, I am husband, father, and grandfather; and to each title is attributed a unique personality based on their distinct relationship to me. They know me differently and, as far as they know me, have different perceptions of who I am. In this way, I incorporate three separate personalities, but they are unified in my one person and are recognized by my family as the man named David Tozar. If this doesn't help, just let it go. Suffice it to say that one united God possesses three distinct personalities.

Why is it important for Jesus to be both God and man? Remember that the trouble started when a man rebelled against the authority and holiness of God. To atone for this rupture, a man was required. But if we go back to the Old Testament book of Leviticus, we find the rules relating to the Hebrew rituals of sacrifice. These sacrifices were to atone for any number of sins committed in all manner of circumstances. There is even one for sins that were not committed consciously. The points we want to focus on are two: one is that the animal offered had to be without blemish, and the second is that the absolution of the particular sin was temporary. As far as a permanent solution is concerned, I don't imagine anyone would think that an animal's blood would be of any real use to anyone.

Now, where was one to find a man without blemish? A man would have to answer for what man had done. As we know, all men are blemished, full of self-centeredness and a tendency to evil. So a mere man could not answer for a divine rupture. And how would a temporary Band-Aid heal a mortal wound anyway? Here's the long and short of it. God sent himself down to be the permanent atonement for all of man's sins. Only he was an acceptable substitute for us, and he was capable of satisfying the debt for all men. God himself came to live among us as a man having willingly divested himself of his divine power while still retaining his divinity in his nature.

Can you even begin to fathom the love that God has for us? In spite of what we deserve—after all, the separation was our doing—he gives us a way back. The only condition is that we accept the way that he has provided. And we need to remember that when we talk of Jesus's death, it is important to recognize the he was not crucified for the things he *did*; in fact, we find none of his detractors arguing about whether he performed miracles or not. Both those who followed him and those who hated him agreed that he did perform them. No, he was crucified for *who* he was.

There is something else to consider. Jesus did not come to make society *right*. He did not come to make sure that good people were put in charge of everything. He did not come to correct social injustice or redistribute wealth. He did not come to cure poverty or crime. He came to fulfill the Father's commands. He lived the kind of life meant for us but that we in our selfish nature are incapable of living. He lived to be the acceptable sacrifice. He lived so that he could exchange his righteousness for our evil. He came to ransom us from our self-imposed bondage. He came to glorify God.

It should be mentioned that many people have described Jesus as a "great teacher," "moral leader," and a "great life example" and are willing to accept him as such but nothing more. But this is the one thing we cannot do. In reality, with a close review of his sayings, we must conclude that he was the boldest liar that ever lived, was completely insane, or he was who he claimed to be. No other options are open to us, and, I believe, no others are intended.

(By the way, this is not the Jesus of Mormonism, the Jehovah's Witnesses, or the Qu'ran. The *real* Jesus is not described in these other faith systems. The name of Jesus is only used to aid them in attaining their own ends.)

Jesus was crucified. This was a form of punishment generally consigned to capital offenders. The type of crucifixion he suffered was a hideous Roman "improvement" of an earlier method of impaling someone on a stake. The prisoner was usually stripped and ruthlessly beaten prior to being attached to a cross. Iron nails about seven inches long were driven through the wrist of each outstretched arm, affixing them to a cross beam. The legs were stretched out, the feet placed one on top of the other, and an additional nail driven through them both. The entire apparatus, the cross, was then lifted and dropped into a hole in the ground and the sufferer left there to die.

Death, however, didn't come quickly. In addition to the agony of the nails, you now had added to it the weight of the body tearing bone, ligament, and cartilage. You had a choice of pushing up with your feet or hanging by your wrists. To give the feet some relief, you would hang, but that would force the lungs to close so that you couldn't breathe. There was a constant pushing up on the feet and hanging from the wrists—up and down, up and down—all the while the nails were grinding bone and flesh. Men could go on like this for days. (We are enamored today with the ruins of ancient Rome, the so-called glory of Rome. Reading histories of these glorious fellows leaves me wondering if there were ever more insidious, bloody-minded people on the globe. And they called the Huns and Mongols barbarians!)

But Jesus's crucifixion was even worse if you can imagine anything more so. In addition to the physical torture, he had to endure the filth of mankind dumped on his shoulders. Should you have the imagination to do so, try to think of something you did in the past that you have felt guilty about ever since. Maybe it wasn't a big thing as the world judges, but it has always bothered you. Add to this the realization that the billions of people that came before you also had such things they were guilty of. As they are piled on top, consider all the truly bad people then and now. I mean, ruthless murderers, rapists, burners of cities, terrorists—all these terrible deeds were done by *people* and they had to be accounted for. All this was placed on the shoulders of the *one* person who had lived a blameless life.

At the end, Jesus did not succumb to his wounds. When the payment for all the sins of mankind had been extracted from him, he surrendered his spirit back into the Father's hands. As he passed away, he says, "It is finished." He did not mean to say that his life of sorrow was ended. What he meant was that the goal was reached; his intended purpose was accomplished. He had satisfied our debt by exchanging his purity for our filth.

Once again, though, there is a caveat. We must own our part of the debt, recognize the need to have it paid, and believe that Jesus paid it. It is imperative that we see ourselves hanging on that cross because that's where we *should* have been and were, in fact, if we accept that Jesus died for us. You and me.

There are other equally important things that Jesus accomplished. On the third day, some women friends of his found the tomb in which he had been enclosed was empty. He had risen from the dead and over a period of time physically showed himself to hundreds of people, most importantly his disciples. Finally, he ascended into heaven, promising to send a comforter to them. This same comforter comes to each of us who believe in Jesus the Christ. This is the Holy

Spirit, which is the third personality of the Trinity, and it is only with his aid that we are able to discern the truth of God's Word. We should not be surprised that materialists consider believers lost in a world of hocus-pocus. They are blind and cannot discern the truth.

Let's take a moment to look at just what is meant by belief. Some call it faith, and that would be fine if that word had not been so maligned over the centuries. Critics like to refer to it as "blind faith" (sort of like believers in materialism—sorry, couldn't resist). Others say that faith is believing in something that isn't true. That certainly isn't what Jesus asked of his disciples when he told them that if they believed in him they would have eternal life. He asked his disciples to look at the evidence. Consider what they had themselves witnessed. As the disciple, John tells us:

> That which was from the beginning, which we have heard, which we have seen with our own eyes, which we have looked upon, and our hands have handled, concerning the Word of life – the life was manifested, and we have seen, and bear witness, and declare to you the eternal life which was with the Father and was manifested to us – that which we have seen and heard we declare to you, that you also may have fellowship with us; and truly our fellowship is with the Father and with His Son Jesus Christ. And these things we write to you that your joy may be full. (1 Jn. 1:1–4)

This is not blind faith or even faith as we generally use the word. This is a promise. A plainer declaration of truth has never been made. The Old Testament is full of instances of people who acted the fool either by relying on their own wisdom or riches or some other material thing. That was pre-Christ, of course, but the apostle Paul knew that man's arrogance would not end even with the risen Christ. Many would continue to refuse God's promises now just as they had throughout history. Just to show how futile our efforts on our own are, he said:

> For the message of the cross is foolishness to those who are perishing, but to us who are being saved it is the power of God. For it is written:

> "I will destroy the wisdom of the wise, and bring to nothing the understanding of the prudent."
>
> Where is the wise? Where is the scribe? Where is the disputer of this age? Has not God made foolish the wisdom of this world? For since, in the wisdom of God, the world through wisdom did not know God, it pleased God through the foolishness of the message preached to save those who believe. For Jews request a sign, and Greeks seek after wisdom; but we preach Christ crucified, to the Jews a stumbling block and to the Greeks foolishness, but to those who are called, both Jews and Greeks, Christ the power of God and the wisdom of God. Because the foolishness of God is wiser than men, and the weakness of God is stronger than men." (1 Cor. 1:18–25)

And again:

> For you see your calling, brethren, that not many wise according to the flesh and not many mighty, not many noble, are called. But God has chosen the foolish things of the world to put to shame the wise, and God has chosen the weak things of the world to put to shame the things which are mighty; and the base things of the world and the things which are despised God has chosen, and the things which are not, to bring to nothing the things that are, that no flesh should glory in His presence. But of Him you are in Christ Jesus, who became for us wisdom from God—and righteousness and sanctification and redemption—that, as it is written, "He who glories, let him glory in the Lord." 1 Cor. 1:26–31

Let's take a moment to tie the two testaments together by uniting them in Jesus. The fundamental reason that many Jews could accept Jesus as the Messiah, (they were the first Christians) besides what they had seen and heard themselves, were the prophecies that were fulfilled by him. To review just a few:

> Born of woman
> Gen. 3:15: "And I will put enmity between you and the woman and between your seed and her Seed: He shall bruise you head, and you shall bruise His heel."

Gal 4:4: "But when the fullness of time had come, God sent forth His Son, born of a woman, born under the law."

Virgin birth

Isa 7:14: "Therefore the Lord Himself will give you a sign: Behold, the virgin shall conceive and bear a Son, and shall call His name Immanuel" (Immanuel—God with us).

Matt. 1:18, 24, 25: "She was found with child of the Holy Spirit.... Then Joseph... did not know her till she had brought forth her firstborn Son. And he called His name Jesus."

The Son of God

Ps. 2:7: "I will declare the decree: The Lord has said to Me, 'You are My Son, today I have begotten you.'"

Matt. 3:17: "And suddenly a voice came from heaven, saying, 'This is My beloved Son, in Whom I am well pleased.'"

Descended from Abraham

Gen. 22:18: "In your seed all the nations of the earth shall be blessed, because you have obeyed My voice."

Matt. 1:1: "The book of the genealogy of Jesus Christ, the Son of David, the Son of Abraham."

Descended of Isaac

Gen. 21:12: "But God said to Abraham, 'Do not let it be displeasing in your sight because of the lad or because of your bondwoman. Whatever Sarah has said to you, listen to her voice; for in Isaac your seed will be called.'"

Lk. 3:23–34: "Now Jesus Himself... the son of Isaac."

From the tribe of Judah

Gen. 49:10: "The scepter shall not depart from Judah, Nor a lawgiver from between his feet, Until Shiloh comes; And to Him shall be the obedience of the People."

Heb: 7:14: "For it is evident that our Lord arose from Judah."

From the line of Jesse

Is. 11:1: "There shall come forth a Rod from the stem of Jesse, and a Branch shall grow out of his roots."

Lk. 3:23–32: "Jesus the son of Jesse."

From the House of David

Jer. 23:5: "'Behold, the days are coming,' says the Lord, 'That I will raise to David a Branch of righteousness; A king shall reign and prosper, and execute judgement and righteousness in the earth.'"

Lk. 23:23–31: "Jesus, the son of David."

Born in Bethlehem

Micah 5:2: "But you, Bethlehem Ephrathah, though you are little among the thousands of Judah, yet out of you shall come forth to Me the One to be Ruler in Israel, whose goings forth are from old, from everlasting."

Matt. 2:1: "Jesus was born in Bethlehem of Judea."

Greeted with gifts

Ps. 72:10: "The kings of Tarshish and of the isles will bring presents; the kings of Sheba and Seba will offer gifts."

Matt. 2:1–11: "Wise men for the East came to Jerusalem . . . and fell down and worshiped Him. And when they opened their treasures, they presented gifts to Him."

The Massacre of the Innocents

Jer. 31:15: "Thus says the Lord: 'A voice was heard in Ramah, lamentation and bitter weeping, Rachel weeping for her children, refusing to be comforted for her Children, because they were no more."

Matt. 2:16: "Then Herod, when he saw that he was deceived by the wise men, was exceedingly angry; and he sent forth and put to death all the male children who were in Bethlehem and in all its districts, from two years old and under, according to the time which he had determined from the wise men."

Called Lord

Ps. 110:1: "The Lord said to my Lord: 'Sit at My right hand, till I make Your enemies Your footstool.'"

Lk. 2:11: "For there is born to you this day in the city of David a Savior, who is Christ the Lord."

Called "God with Us"

Is. 7:14 "Therefore the Lord Himself will give you a sign: Behold, a virgin shall conceive and bear a Son, and shall call His name Immanuel."

Matt. 1:23: "'Behold, the virgin shall be with child, and bear a Son, and they shall call His name Immanuel,' which is translated, 'God with us.'"

A Prophet

Deut. 18:18: "I will raise up for them a Prophet like you from among their brethren, and will put My words in His mouth, and He shall speak to them all that I command Him."

Matt. 21:11: "So the multitude said, 'This is Jesus, the prophet form Nazareth of Galilee.'"

A Priest

Ps. 110:4 "The Lord has sworn and will not relent, 'You are a priest forever according to the order of Melchizedek.'"

Heb 3:1: "Therefore, holy brethren, partakers of the heavenly calling, consider the Apostle and High Priest of our confession, Christ Jesus."

A Judge

Is. 33:22: "For the Lord is our Judge, the Lord is our Lawgiver, the Lord is our King; He will Save us."

Jn. 5:30: "I can Myself do nothing. As I hear, I judge; and My judgement is righteous, because I do not seek My own will but the will of the Father who sent me."

A King

Ps. 2:6: "Yet if have set My King on My holy hill of Zion."

Matt. 27:37: "And they put up over His head the accusation written against Him: THIS IS THE KING OF THE JEWS."

Anointed by the Holy Spirit

Is. 11:2: "The Spirit of the Lord shall rest upon Him, the Spirit of wisdom and understanding, the Spirit of counsel and might, the Spirit of knowledge and the fear of the Lord."

Matt. 3:16, 17: "When He had been baptized, Jesus came up immediately from the water; and behold, the heavens were opened to Him, and He saw the Spirit of God descending like a dove and alighting upon Him. And suddenly, a voice came from heaven, saying, 'This is My beloved Son, in whom I am well pleased.'"

Zealous for the Father

Ps. 69:9: "Because zeal for Your house has eaten me up, and the reproaches of those who reproach You have fallen on me."

Jn. 2:15, 16: "When He had made a whip of cords, He drove them all out of the temple . . . and He said, 'Take these things away! Do not make My Father's house a house of merchandise!'"

Preceded by John

Is. 40:2: "A voice of one crying in the wilderness: 'Prepare the way of the Lord; Make straight in the desert a highway for our God."

Matt. 3:1: "John the Baptist came preaching in the wilderness of Judea, and saying, 'Repent, for the kingdom of heaven is at hand!'"

Will Perform Miracles

Is. 35:5, 6: "Then the eyes of the blind will be opened, and the ears of the deaf unstopped. Then the lame will leap like a deer, and the tongue of the dumb will shout for joy."

Matt. 9:35: "And Jesus was going about all the cities and villages, teaching in the synagogues, and proclaiming the gospel of the kingdom, and healing every kind of disease and every kind of sickness."

Speak in Parables

Ps. 78:2: "I will open my mouth in a parable; I will utter dark sayings of old."

Matt. 13:34: "All these things Jesus spoke to the multitude in parables; and without a parable He did not speak to them."

Would enter Jerusalem on a donkey

Zech. 9:9: "Rejoice greatly, O daughter of Zion! Shout, O daughter of Jerusalem! Behold, you King is coming to you; He is just and having salvation, humble, lowly and riding on a donkey, a colt, the foal of a donkey."

Lk. 19:35–37: "And they brought him to Jesus. And they threw their own clothes on the colt, and they set Jesus on him. And as He went, many spread their clothes on the road. Then, as He was now drawing neat the descent of the Mount of Olives."

Was also coming for the Gentiles

Is. 60:3: "And Gentiles shall come to your light, and kings to the brightness of your rising."

Acts 13:47: "For so the Lord has commanded us, 'I have set you as a light to the Gentiles, that you should be for salvation to the ends of the earth.' Now when the Gentiles heard this, they were glad and glorified the word of the Lord."

One of his would betray him

Ps. 41:9: "Even my familiar friend in whom I trusted, who ate my bread, has lifted up his heel against me?"

Matt. 10:4: "Judas Iscariot, who also betrayed Him."

Betrayed for thirty pieces of silver

Zech. 11:12: "Then I said to them, 'If it is agreeable to you, give me my wages; and if not, refrain.' So they weighed out for my wages thirty pieces of silver."

Matt. 26:15: "'What are you willing to give me if I deliver Him to you?' And they counted out to him thirty pieces of silver."

The disciples would desert
Zech. 13:7: "Strike the Shepherd, and the sheep will be scattered."
Mk. 14:50: "Then they all forsook Him and fled."

Would rise from the grave
Ps 16:10 "For You will not leave my soul in Sheol; nor will You allow Your Holy One to see corruption."
Matt 28:6 "He is not here, for He has risen, just as He said. Come, see the place where He was lying."

These fulfilled prophecies go on and on, but I believe this sampling will suffice. Jesus himself referenced Old Testament prophecies as pertaining specifically to himself. Two of the many examples are the following:

Jn. 5:39, 40, 46, 47: "You search the Scriptures, for in them you think you have eternal life; and these are they which testify of Me. But you are not willing to come to Me that you may have life. For if you believed Moses, you would believe Me; for he wrote about Me. But if you do not believe his writings, how will you believe My words?"

Lk. 22:44: "Then he said to them, 'These are the words which I spoke to you while I was still with you, that all things must be fulfilled which were written in the Law of Moses and the Prophets and the Psalms concerning Me.'"

The resurrection of Christ is the bedrock on which Christianity rests. If this event has not occurred, in fact, then all the faith in the world will not save any of us. Of the major religions of the world, the followers of Christ are the only ones who claim a physical resurrection from the dead, an ascension into heaven, and an intercessor who opens the door to the Father on our behalf. Abraham, the father of Judaism, died in about 1900 BC. He is, well, still dead. Buddha, in the Mahaparinimmana Sutta, says that he died in such a way that he went "with that utter passing away in which nothing whatever remains behind." I believe him. He too is still dead. Muhammad died on June 8, 632, at the age of sixty-one at Medina, where every year his tomb is visited by tens of thousands of his followers. His bones are still there.

If the resurrection was not firmly believed by the apostles as having taken place, if they did not believe that they had physically seen him alive, Christianity could never have come about. The crucifixion would have been the bitter end of a sad career. Furthermore, Jesus said on more than one occasion that he would be put to death in Jerusalem and on the third day rise. Peter's famous sermon during Pentecost is wholly and completely founded on the resurrection. Beginning in Acts 2:17, he provides some background from the Old Testament prophet Joel. Joel was speaking of the second coming of Christ, but Peter used it as a sort of sample of what will happen during the millennial kingdom when all people will have the Spirit poured out on them. But starting in verse 22, he says:

> Men of Israel, hear these words: Jesus of Nazareth, a Man attested by God to you by miracles, wonders and signs which God did through Him in our midst, as you yourselves know—Him, being delivered by the determined purpose and foreknowledge of God, you have taken by lawless hands, have crucified, and put to death; whom God raised up, having loosed the pains of death, because it was not possible that He should be held by it.

Picking up in verse 29

> Men and brethren, let me speak freely to you of the patriarch David, that he is both dead and buried, and his tomb is with us to this day. Therefore, being a prophet, and knowing that God had sworn with an oath to him that the fruit of his body, according to the flesh, He would raise up the Christ to sit on his throne, he, foreseeing this, spoke concerning the resurrection of the Christ, that His soul was not left in Hades, nor did His flesh see corruption, we are all witnesses. Therefore, being exalted to the right hand of God, and having received from the Father the promise of the Holy Spirit, He poured out this which you now see and hear.

According to authors Ronald Tacelli and Peter Kreeft:

> The resurrection also sharply distinguishes Jesus from all other religious founders. The bones of Abraham and Muhammad and Buddha and Confucius and Lao-Tzy and Zoroaster are still here on earth. Jesus tomb is empty.
>
> The existential consequences of the resurrection are incomparable. It is the concrete, factual, empirical proof that: life has hope and meaning; "love is stronger than death"; goodness and power are ultimately allies, not enemies; life wins in the end; God has touched us right here where we are and has defeated our last enemy; we are not cosmic orphans, as our modern secular worldview would make us. (Kreeft, *HC*, 177)

Not only is the resurrection of critical importance but the fact that Jesus spoke of it prior to the event proved that he was who he claimed to be. Wilbur Smith puts it this way:

> It was this same Jesus, the Christ who, among many other remarkable things, said and repeated something which, proceeding from any other being would have condemned him at once as either a bloated egoist or a dangerously unbalanced person. That Jesus said He was going up to Jerusalem to die is not so remarkable, though all the details He gave about that death, weeks and months before He died, are together a prophetic phenomenon. But when He said that He himself *would rise again from the dead*, the third day after He was crucified, He said something that only a fool would dare say, if he expected longer the devotion of any disciples—unless He was sure He was going to rise. No founder of any world religion known to men ever dared say such a thing as that! (Smith, *GCWC*, 10–11)

Let's look at the information these folks have accessed to arrive at these conclusions. Arguments presented by those who claim that Jesus was merely in a swoon or that he was *nearly* dead simply do not take into account the very thorough way in which the Roman soldier approached his job. There was simply no way these men were going to take anyone down before he was good and dead. Dead as a mackerel. As if crucifixion itself was not bad enough, the soldiers who had drawn this detail broke the legs of the two thieves to hurry their departure from this earth in deference to the Jewish Passover. This was their way of showing mercy! I think they were just happy to get off work early. Jesus, on the other hand, received a spear thrust under his ribs that penetrated his heart to make sure that he was good and most thoroughly dead.

There are basically four facts that were reported that all historians agree on regardless of their individual worldview. There have been over 1,400 academic sources published in English, French, and German since 1975 that support this statement. These four facts are:

1. Jesus was crucified on a Friday on a cross constructed by Romans, died, and was buried in a tomb.
2. On the ensuing Sunday, said tomb was empty.
3. At great personal risk, numerous witnesses testified to having seen him alive after he had died.
4. Jesus's brother James and the archpersecutor of Christians Saul were so convinced that they had seen the living Jesus they preferred to be killed rather than recant their story.

The recorded sightings of Jesus after he had died are as follows:

To Mary Magdalene	Jn. 20:11–18
To the other women	Matt. 28:8–10
To Peter	Lk. 24:34
To ten disciples	Lk. 24:36–43, Jn. 20:19–25
To the eleven including Thomas	Jn. 20:26–29
At his ascension	Lk. 24:50–53, Acts 1:4–12
To the disciples on the Emmaus Road	Lk. 24:13–35
In Galilee	Matt. 28:16–20, Jn. 21:1–24
To five hundred people	1 Cor. 15:6
To James and the apostles	1 Cor. 15:6
To Paul on the road to Damascus	Acts 9:1–6, 18:9, 10, 22:1–8, 26:12–18, 1 Cor. 15:8

Why do people reject out of hand the resurrection? If the level of eyewitness testimony available in this case was available in any other, you'd be considered crazy if you denied it. It's simple: they are stuck in what they call an intellectual difficulty. This difficulty is derived solely from the twenty-first-century claim that any fool knows that there are no such things as miracles; it's been proven! The argument goes something like this: "I think there is a lot of good stuff in the Bible but I just can't intellectually wrap my mind around a dead man rising from the grave." Well, let's see, you mean you *can* get your mind around the universe coming to be without a cause? You *can* get your mind around a catfish becoming a camel? Is it reasonable to you to believe that dead stuff turned into live stuff? If so, why is it so hard to believe that a dead man came back to life? Let me be clear: intellectualism has nothing to do with their difficulties. To be intellectually honest means that you follow the evidence down whatever road it takes you regardless of whether it's the road you expected or preferred to be on.

I believe that those who propose that the sightings of a risen Jesus were imagined pose a legitimate question. Our own experience and the experience of others tells us that people imagine things on a regular basis (some more than others). But how can you apply this scenario to this situation? *Many* saw Jesus and these sightings were in different locations. These folks could not have imagined the same thing. Besides, what did they hope to gain by such pronouncements if they were not true? To be officially branded crazy? To be killed by those who had Jesus killed? To be totally ostracized by their families, friends, and the very faith they were brought up in? To ceaselessly be crying out nonsense that every other person knew to be a lie? Since I know something of human nature, it strikes me that this requires much more of faith than believing that what they reported regardless of its unusual nature was true.

That's another thing that I think throws people—the unusual nature of the report. It is not only unusual but is one of a kind. But why should that be a stumbling block to the intelligence solely for it not being a more common occurrence? How much attention would you pay to people rising from the dead if your community was overcrowded because of all the empty graveyards? If you consider God to be a common sort of being, I have nothing more to say to convince you otherwise. Once we accept that God is who he has revealed Himself to be, we should laugh and enjoy the wondrous things he has done.

Chapter Ten

THE FOUNDING OF THE CHURCH

So I had accepted the Bible as an accurate portrayal of history and the death, burial, and resurrection of Christ. Now what? The church of Christ has been around a long time, and things are still a mess (and have been since the garden was closed to the public). Why wasn't the gospel having more of an impact through this church that is reputed to be the body of Christ on earth? Before I committed to belonging to something, not being much of a joiner, I had to take a hard look at this church. Let's face it, the church has had its own problems over the past two thousand years. So how, after such a literally divine beginning, did so much confusion and nonsense enter in to so simple a concept? I decided to take time and study the history of the church and come to some understanding of how so many varieties came about.

It is important to point out that all the original followers of the Way (as Christians were first known) were Jews. There had been some outreach to Gentiles, most notably with the woman at the well in Samaria, but there was no question of conversion at that point simply because Jesus had not been crucified, let alone ascended to heaven. There was no church. There is, however, mentioned in Acts

6:1 the subject of Hellenists. It says: "For the Hellenists murmured against the Hebrews because their widows were neglected in the daily distribution." The people Luke is referring to here as Hellenists are not Gentiles (Greeks) but the more cosmopolitan Jews who had taken on some of the forms of Hellenization following the conquest of Palestine by Alexander. It was, in fact, just how Gentiles were to be introduced into the faith (or *if* they were) that the first major controversy came about.

The Jews, by the time of Christ, had long been scattered throughout the Roman empire and beyond. As evidence of this, in Egypt they had built two temples centuries earlier and in Rome itself an entire section of the city was Jewish. This scattering of the twelve tribes was known as the Diaspora or Dispersion. Strong ties to the ancient homeland still existed, but it was much easier for the people thus dispersed to become Hellenized. These settlements proved to be one of the main conduits by which the message of Christ through his disciples was spread. In addition, these Jews provided, while probably unintentionally, one of the greatest resources used by Christians to lend historical credence to their message. This resource was a Greek translation of the Old Testament. It was called the Septuagint or simply LXX (70—taken from the number of scribes needed to translate it, or so it was said). This had been produced simply because most of the resettled Jews had long since forgotten their native tongue. After Alexander, Greek became the primary language in most areas he had conquered.

The reason this Greek Old Testament was such a powerful tool is that Jews from Palestine could use the very "Bible" of the Hellenized Jews to point out all the writings describing and prophesying the coming of Messiah, all of which Jesus fulfilled. Further, many Jews in these areas had already had news of the Way. (Christians were originally described as 'followers of the way.') Every major city in the Roman empire had their resettled Jewish population (even some towns not so major). Travel because of commerce and relocation had been commonplace for centuries by both land and sea. Rome had built a marvelous network of roads throughout the empire, and caravans still traveled over the ancient routes from east to west and back again. The Mediterranean saw fleets of ships running all over its length and breadth, so much so that piracy was occasionally a problem. In this fashion, news of Jesus traveled much like what would become known as the "bamboo telegraph" in the Philippines during World War II.

Now, let us return to Jerusalem. In chapter 2 of the book of Acts, the "comforter" that Jesus had promised arrived. Peter's sermon on Pentecost is eloquent and convincing to the group of Jews surrounding him; we are told that there were three thousand of them. The Sanhedrin, those who had arrested Jesus and taken him to trial, shortly afterward arrested both Peter and John for preaching the "good news" of Jesus Christ. This, of course, didn't stop them. Again, it must be mentioned that these two men (and hundreds of others) did not desist when ordered by men fully capable of having them killed. Why? The only reasonable answer is that they were fully convinced in the truth of what they were preaching.

As mentioned, the first controversy in the church was centered around the acceptance of Gentiles/Greeks (all non-Jews were thus considered) into the faith. The early Jewish converts did not consider themselves followers of a new religion. They were still Jews. Their new faith did not deny Judaism but accepted Jesus as the fulfillment of the long-anticipated Messiah of prophecy. This was the natural, to them, extension of their Jewishness. Jews had a leg-up in that they had nursed with the recitation of Old Testament verses sounding them to sleep. Jewish schools were centered around Old Testament teachings along with the rules and rituals that had been piled on during the ensuing years. They understood that there was one God, Yahweh, that man had fallen from grace and all that had occurred since. Gentiles had no such advantage. They had grown up with more gods and goddesses than you could shovel across the barnyard in a year's time.

The controversy stemmed from this disparity in backgrounds. Since Gentiles did not have the Old Testament as a foundation, Jesus could very easily become assimilated into the already-overstuffed basket of mythological beings. Because of the reality and importance of Jesus's sacrifice, the early Christians refused to allow it to be cheapened because would-be converts lacked a complete understanding of what they were professing to believe. So many in the church at Jerusalem insisted that these Gentile-converted followers of the Way had to become Jews first. They needed a thorough indoctrination into Old Testament scripture, circumcision, rules and rituals, etc. How this would take shape was being debated while other events were transpiring.

Stephen was the first martyr of the faith (Acts 7); and standing in the crowd, holding the coats of the murderers, was a gentleman named Saul who hailed from Tarsus. While the argument surrounding how to accept Gentiles was going on among Christian Jews, aider and abettor of murderers, Saul, was on his way to the city of Damascus to secure more Christians for trial back in Jerusalem. He was zealous if nothing else. One thing that helps to understand his fury in persecuting this new "sect" is the realization that he was a Pharisee. The Pharisees, remember, were interested in retaining control through the continuance of the religious *system*. The system kept those in power who had it and kept those of lesser station in their place. But something strange and unexpected happened to Saul while on the road.

> Then Saul, still breathing threats and murder against the disciples of the Lord, went to the high priest and asked letters from him to the synagogues of Damascus, so that if he found any who were of the Way, whether men or women, he might bring them bound to Jerusalem. As he journeyed he came near Damascus, and suddenly a light shone around him from heaven. Then he fell to the ground, and heard a voice saying to him, "Saul, Saul, why are you persecuting Me?" And he said, "Who are you, Lord?" Then the Lord said, "I am Jesus, whom you are persecuting. It is hard for you to kick against the goads." So, he, trembling and astonished, said, "Lord, what do You want me to do?" Then the Lord said to him, "Arise and go into the city, and you will be told what you must do." (Acts 9:1–9)

So Saul became Paul, and his story thereafter is the story of the greatest evangelist in history. More people have come to faith in Christ through his writings than anyone else whether Jew or Gentile. This is yet another example of God fulfilling his purposes through unworthy and broken men. Paul started as the terror of Christians and became a teacher of the convert and possessor of the most profound love for he whom he had persecuted. In microcosm, this is the tale of billions of changed lives through Jesus, including mine.

Now, back to our controversy. Peter too had preached to the Gentiles, and he had personally witnessed them being possessed of the Holy Spirit—the same comforter/teacher that the Jewish converts received. At the same time, Paul was strengthening the Gentile churches in his area. But this insistence on the "Judaizing" of non-Jewish converts was reaching them and causing considerable consternation. In Jerusalem, a council had been called so all the apostles and elders in the church gathered there to consider the question in earnest. It was Peter who said:

> "Men and brethren, you know that a good while ago God chose among us, that by my mouth the Gentiles should hear the word of the gospel and believe. So God, who knows the heart, acknowledged them by giving them the Holy Spirit, just as He did to us, and made no distinction between us and them, purifying their hearts by faith. Now, therefore, why do you test God by putting a yoke on the neck of the disciples which neither our fathers nor we were able to bear? But we believe that through the grace of the Lord Jesus Christ we shall be saved in the same manner as they." Then all the multitude kept silent and listened to Barnabas and Paul declaring how many miracles and wonders God had worked through them among the Gentiles. (Acts 15:7b–12)

In the end, it was agreed that the Gentiles should be welcomed into the church with the admonition to refrain from offering to idols, from blood, from things strangled and sexual immorality. This controversy would rear up now and then, but as far as the church leadership was concerned, it was settled. More would arise, however, as the number of believers continued to grow and outsiders would attempt to co-opt the growing faith to their own interests.

Peter and John were two of the "pillars" of the budding church as mentioned by Paul in Galatians 2:9. The third was not an apostle but James, a half brother of Jesus. James, having grown up in the same house as Jesus, had no extraordinary opinion of him that we know of, but upon being visited by his risen half brother, he saw the light (so to speak) and became a believer (1 Cor. 15:7). In fact, brother James became the leader of the church in Jerusalem and was ultimately martyred for his faith.

From this beginning, the church continued to spread in all directions. As mentioned previously, all the original Christians were Jews and were located at Jerusalem and some of them were of the Hellenized sort. Since these more urbane Jews mixed more readily with Gentiles (Greeks), they provided a bridge to these folks that eased the spreading of the good news. Also, Philip founded a church in Samaria, a previously taboo area of "untouchables" because of their being largely a mixed breed of Jews and transplanted Assyrians.

Over the next couple of centuries, the church would experience intense periods of persecution and relative periods of calm as the Roman empire had other more pressing concerns. Despite every effort to exterminate the faith by the emperors, it continued to grow. Finally, in AD 322, Constantine became the sole ruler of the empire, persecutions stopped, and Christianity was welcomed largely through the conversion of the emperor himself. It is interesting to observe that, though the persecutions were ended and the average Christian rejoiced, Constantine, perhaps inadvertently, would be the cause of future problems. Contrary to what many people believe, Constantine did not make Christianity the state religion (that would come later) and he had his detractors. Most of these detractors were to be found in Rome, so the emperor decided to build a "New Rome" in the eastern portion of the empire. Thus, the city of Constantinople came to be built, a new center of the faith was founded, all while the church in Rome also grew in stature and significance.

From Constantine forward, the state had a vested interest in theological questions because it was hoped that Christianity would help to hold the empire together and avoid the civil wars that had proven so destructive in recent decades. In the early days of the church, controversies were settled by long debate and searching of the scriptures until a consensus was eventually reached. But once the emperors became involved, who were not interested in prolonged discussions and debates, they simply settled things based on imperial authority. As a result, more theological questions were taken to the emperor for a solution rather than the church elders. The old forms of debate were soon replaced by newer forms of political intrigue.

While the empire was being consolidated under Constantine, another serious controversy arose to shake the church. Man, trying to read between the lines of scripture and thinking themselves able to understand God, tried to explain the mystery of the Trinity in their own way and from these efforts arose the Arian controversy. The germ of the situation was rooted in the writings of some early church fathers and theological scholars such as Justin Martyr, Clement of Alexandria, and Origen.

The argument centered around the way in which some churchmen were coming to view the nature of God himself. Initially, since Christians seeking to spread the good news had no visible God to reference, like pagans had with the sun, stars, or a human representation, it was difficult for the pagan to visualize what they were talking about. So to fill this gap and to allow the Gentile to "see" the invisible, they co-opted the classical philosopher's notion of a supreme being ruling over the entire cosmos. This certainly aided in the acceptance of Christianity among these people, especially the educated, but was also fraught with misunderstanding and danger to the basic tenants of the faith.

The Christians who adopted this method of linking their faith with philosophy considered it the best (read: most expedient) way of articulating the Way. Rather than relying on the literal reading of scripture and educating them in the Old Testament, they started resorting to allegory. Logos, the Greek for *word* or *reason*, was used as a separate entity, sort of an associate or spokesperson for God. However, this concept of God as an uninterested ruler over all effectively removed the Christian from being able to have a personal relationship with him. This negated the very message of salvation, the fundamental purpose of which was to renew this relationship and for which Jesus sacrificed himself. The writing of Justin, Origen, et. al, sought to integrate philosophy in this manner: God the Father was unable to reveal anything of himself, but the Logos or Word of God could and was capable of relationship with people. This became widely taught and accepted in the eastern portions of the empire, which was centered at the time in Alexandria, Egypt.

The bishop of Alexandria, Alexander, had some disputes with a popular preacher (called a presbyter) named Arius. Up to this time they had been relatively minor affairs. Things became heated when

Arius started insisting that God the Father and the Logos were not only distinct but coeternal. (Philosophically, this is a nonstarter. There cannot be two eternal beings.) Alexander stood on the biblical doctrine of one God having three personalities—the Trinity. Taking Jesus as the Word (as John had referred to him) as one of those three personalities, he could not look at any separation of the Godhead as less than blasphemy. The crux of the problem was that Arius considered Jesus a created being (which notion negated him as being coeternal in any event) and therefore denied the Trinity. If this were so, then Jesus died for nothing, was unable to satisfy our debt to God, we are still in our sins, and Christianity is meaningless. There were those who backed Arius though they were few in number.

While this kettle was beginning to boil, Constantine had completed his consolidation of the empire under his sole rule. As problems among the Christians were not helpful in his desire to unify the empire, he decided to step in. His first effort was to send a mediator to see if the two parties would settle things amicably. This having failed, he called a council of bishops to meet in Nicea in the year 325. While an exact count is not known, it is estimated that over three hundred bishops and their aides attended with many of them still having visible scars earned from the recent persecutions. From all parts of the empire, they came with the emperor defraying their expenses. It became known as the First Ecumenical Council.

Among other topics, they discussed church legislation. Christians were now free to worship, and the ranks were swelling, so biblically based unity needed to be articulated. In addition, the readmission of lapsed church members (lapsed because they chose not to be martyred during the persecutions) and the election/ordination process for presbyters and bishops had also to be settled upon. The main topic was, however, the controversy created by Arius. There were only a small group present that were convinced that the view expounded by Arius was correct. They thought that if the logic of the argument were laid out in plain language, then all the other attendees would come to agree with them. In this, they were badly mistaken. The council decided it was time to put the basics of their belief as Christians into a statement, which, in the event, flatly denied Arianism that converts would need to subscribe to. This statement of faith said:

> We believe in one God, the Father Almighty, maker of all things visible and invisible.
>
> And in the Lord Jesus Christ, the Son of God, the only-begotten of the Father, that is, from the substance of the Father, God of God, light of light, true God of true God, begotten, not made, of one substance [homoousios] with the Father, through whom all things were made, both in heaven and on earth, who for us humans and for our salvation descended and became incarnate, becoming human, suffered and rose again on the third day, ascended to the heavens, and will come to judge the living and the dead.
>
> And in the Holy Spirit.
>
> But those who say that there was when He was not, and the before being begotten He was not, or the He came from that which is not, or that the Son of God is of a different substance [hypostasis] or essence [ousia], or the He is created, or mutable, these the Catholic church anathematizes.

With some alterations and the dropping of the anathemas, this formed the basis for what became known as the Nicene Creed.

This, however, was not to be the end of the teachings of Arius. Over the years, Arianism became increasingly abstract and more deeply drawn into philosophical technicalities and more distant from Christ as the source of salvation for man. The short story is that the eastern church came to be more closely aligned with the Arian view and the church at Rome the defenders of the "true faith." Alexandria was still the eastern center of culture and learning, so it was rather easy, unfortunately, for these "educated men" to infiltrate the church.

There were many great theologians during this period, and I haven't the space to review them all here. One of them stands above the rest, so I would be remiss if I didn't at least briefly mention him. Indeed, he is the most quoted teacher after the apostles and is considered the representative of the western church. Augustine of Hippo was born in 364 to a pagan Roman father and an ardent Christian mother. Early on, he was recognized as a gifted child. His parents sent him off to school at Madaura, but he had to return when their resources ran low. During this hiatus from his studies, he took to carousing with his companions and sampled what temptations his hometown held for a healthy young man.

When he was seventeen years old, a patron named Romanianus stepped forward and enabled him to return to school, this time in Carthage. That ancient city had long been recognized as the political, economic, and cultural center of Romanized Africa. Here he studied hard but also continued to play hard. During this period in his life, he managed to sire a son named Adeodatus with an unnamed woman who would remain with him for many years.

Augustine's field of study was rhetoric, which comprised the art of speaking and writing eloquently. Truth was not a requirement for turning out a fine orator. But one of the authors he read and admired was Cicero, the pride of Rome. Cicero was not only a great speaker but a philosopher as well, so truth was held at a premium by him, and this would rub off on Augustine.

Through many twists and turns, he finally arrived at his mother's faith—perhaps through her prayers. It was not in his nature to approach something so important as a life change halfheartedly, and he suffered many inner torments along the way to faith. Augustine admitted to being torn between the allure of the world and the righteousness of God. When he finally gave all of himself to the faith, he gathered up his mother and son and headed to North Africa where he desired to take up a life devoted to study and contemplation. Along the way, his mother died; and shortly after his arrival in Tagaste, his son died as well.

Augustine settled in Cassiciacum with a few friends. Here they lived a quiet, orderly life, denying themselves any comforts that they didn't consider absolutely necessary. While residing there, he penned his first Christian works though they had some neoplatonist philosophy mixed in. The works did have the effect, perhaps unknown to him, of spreading his name and producing a degree of fame. His quiet life of the studious semi-recluse was soon to end.

While visiting the town of Hippo in 391, he attended a church whose bishop was named Valerius. The bishop, having spotted Augustine in the crowd, preached a sermon about how God always provides for the flock a shepherd when one is needed. Quite as if on cue, the congregation elected Augustine, much against his wishes, to serve alongside Valerius. Not wanting to lose his prize, within four years Valerius named the reluctant Augustine co-bishop, apparently unaware that this wasn't allowed. Shortly thereafter, Valerius passed away and Augustine became bishop of the church at Hippo.

Augustine wrote two books of particular significance: his *Confessions* and *The City of God*. The *Confessions* are autobiographical in style and described his turbulent jousting with God while being led to the faith. *The City of God* is a remarkable and insightful depiction of two cities. The one city is of God and built, or founded, on his love. The other city is an earthly one whose foundation is also love: man's love for himself. While these two cities at times interact, ultimately there are irreconcilable differences between the two. In the end, God's proves to be eternal while man's turns to dust. It is a fascinating read.

Ashes

In 410, Alaric, king of the Goths, took and sacked Rome. Many churches that had once flourished ceased to exist. The church in Constantinople, however, would continue largely unmolested for another thousand years. In the west, the empire fractured into a great number of cities, princedoms, and petty kingdoms. The Germanic tribes of Goths, Vandals, and Franks also carved out domains from the lands they had conquered. This period would become known to us as the Middle Ages. Despite the destruction, the invaders would assimilate themselves, come to embrace the religion of the people they had conquered, and the church would survive and grow—but changed. Two things were of historic significance in this redevelopment of the church. One of these was the rise of monasticism, and the other was the establishment of the papacy. This second change would eventually be the cause of bloodshed and destruction but also the unwilling means of a new birth of Christian freedom.

Europe was deluged by Germanic and other tribes, and there was no longer any empire worth speaking of to impede them. Lombards, Saxons, Angles, Franks, Bretons, Burgundians, Ostrogoths, Visigoths, and others all scrambled for space. The Vandals had conquered Northern Africa all the way eastward to the border of Egypt and would cross over to conquer Sicily, Corsica, and Sardinia. Marching to Rome, they tore up the city in such a manner as to make old Alaric smile. The Vandal conquest of North Africa was a particular disaster, for there were Arian holdovers there, and the end result was to push them back into contact with the church at large.

In the east Huns and Gepidae pressed on the empire's northern borders, Persians were pressing on its eastern frontier, and the aforementioned Vandals were encroaching on the west. In the eighth century, the eastern empire would fall to the Muslims, but the church was already in disarray because of the influx of so many disparate belief systems. The west ignored the cries for help from the other half of the old empire. A remnant in the east would survive to carry the banner of Christ and identified itself as the Eastern Orthodox Church.

Back in the west, Christianity, historically a pacific religion, had found its way into the army. This had originally occurred with converted soldiers in the old Roman legions. With the rise of Islam, militancy was becoming more acceptable and Christian soldiers came to see themselves as the defenders of both church and state as these two entities had become more closely aligned. This combination of militarism and faith would wreak havoc both at home and abroad over the next several centuries. Popes would bounce between using them for their own purposes and cursing their very existence when they became unallied and left to themselves, rampaged throughout Europe.

Amid the chaos prevailing in the west, on Christmas Day, 800, Charles, king of the Franks, was crowned emperor of the renewed Roman empire formed with the backing of the church. He would be called Charlemagne. But prior to his coronation, he had been busy consolidating his position particularly in Britain and what would become France. In campaigns against the Saxons and their Frisian allies, all were slaughtered who proved resistant to the faith and the rest were baptized. I'm not sure this is what Christ had in mind, but so it happened. Worse would follow.

Despite this rather firm treatment of his enemies, once emperor, he proved to be a man who admired learning. Though uneducated himself, he went to great lengths to revive the centers of learning that he inherited. Alcuin of York was called to his court to reintroduce the knowledge that had been preserved in British and Irish monasteries and largely forgotten on the continent. In addition, Benedict, who had previously abandoned the court to become a monk, was recalled to invigorate the royal abbey at Aniane, which would be Charlemagne's model for the operation of all future monasteries. Theodulf, a man known to be pious as well as learned, was called in and given a mandate to create a school in every church that was to be open to rich and poor alike. One potential problem, however, can be readily seen in all this. Charlemagne considered himself the arbiter of all things both temporal and theological. While he did not dictate doctrine, he did have a hand in selecting who would.

Upon Charlemagne's death, his son Louis (called the Pious) ascended the throne as emperor. In 817, he called an imperial diet (council) and ordered that all monasteries be reorganized to follow protocol as established by the aforementioned Benedict of Ariane. Additionally, bishops and the lower clergy were forbidden to wear jewels or ornate clothing. Another directive as a result of this diet was a proclamation requiring a tithe from everyone, with two-thirds of the money collected to be distributed to the poor. Finally, it was decided to reenact the previously discontinued practice of allowing the people and lower clergy to elect bishops.

While this was happening in Europe, Islam was busy consolidating their conquests. From the northern rim of Africa, they had crossed over to Spain and gained enormous holdings there. An unexpected repercussion of this was to make coinage a pretty rare commodity in Europe. The result of this was that the only method of reward available to emperors, kings, princes, and such was the granting to those they wanted to advance holdings in land. A barter economy ruled. This parsing of the land initiated feudalism. Originally, the land was granted for the life of the individual to whom it had been granted, but these grants soon became hereditary. Every such landholder was beholden to someone both above and below him. To complicate things further, men often held land granted to them by multiple people, so politically it was easy for them to claim conflict of interest. Again, Europe began to fracture.

The societal conditions also had an impact on the church. Charlemagne had placed the papacy in a rather awkward position. On one side, the popes obviously had some power as it was they who put the crown on the heads of emperors. In opposition to this is the fact that Rome itself was often in a state of disarray. The emperors could not control their own capital any more than the church could. In the mid-800s a set of documents called the False Decretals, which claimed to be ancient, were produced, which sought to grant to the pope extraordinary authority. (These documents, more likely than not, were forged.) People at the time took them as authentic; and the pope, Nicholas I, bid the powerful to stop their warring, which they often seemed to participate in for simple lack of anything better to do. He could do this through mediation and, should that fail, by bringing up the big gun of excommunication. Of course, these petty wars were anything but enjoyable for the peasant as the result for them was the burning of their crops and villages and often the raping of their women and murder of their children. But during the reign of Pope John VIII, all this newfound papal authority was drained away.

While attempting to respond to the Muslim invasion, he called upon Lothair II, king of Lorraine, for aid who promptly declined the request. While in Monte Cassino, Lothair II attended communion during which John VIII unwisely cursed both the king and his court. Poor Pope John was thereafter poisoned in his own palace; and it was rumored that when he took too long to die, the poisoner, apparently a steward, split his head with a mallet. After this, no one seemed to have a problem disposing of popes.

The problems that ultimately caused a permanent rupture in the church became firmly rooted at the dawning of the tenth century. Many of the authorities, both secular and church, were looking for the world to end at the close of the millennia. This, of course, would bring about the terrible day of judgment; and these authorities, both secular and church, had reason to be fearful of that day of squaring their accounts with God. Many estate holders, seeking to purchase some relief from purgatory, vested vast tracts of land to the church. As the decades wore on, magistrates neglected their business and society became demoralized. The church unfortunately did not prove to be a source of encouragement, it being just as demoralized. The deepest malaise landed, perhaps, in Rome itself.

The papacy, which had been struggling with a kind of schizophrenia for not a little while, completely succumbed to the darker side of its nature. Antichrist was able to enter Rome through the front door and don papal robes. Charlemagne had provided a strong hand during his reign as emperor, but he had not had time to lay a solid foundation to sustain his discipline and laws. His successors lacked his motivation, strength, and energy and largely sank into lives of indolence and depravity. The contempt in which these men were held is exemplified by the epithets by which they came to be known: the Fat, the Stammerer, the Bald, the Child, the Simple, and the Lazy. It was unfortunate, to say the least, that while under the leadership of these men, Europe faced yet another incursion.

The Vikings, or Norsemen (Northmen), like lice, infested the coastline of Germany and France. They were pagans who had swept down from Denmark, Sweden, and Norway like a tempest. Following the rivers and their tributaries, they reached into the heartland of Europe, sacking cities, raping, and pillaging along their way. When these lands offered nothing further to plunder, they turned their attention toward England. The Vikings would continue to be the nightmare of Europe for over one hundred years. But they were not the only problem. The Saracens (who had supplanted the Vandals) had crossed over to Sicily and southern Italy from North Africa, and the Magyars were pressing Germany and Italy from the east. Moreover, our fun-loving landowners enjoyed nothing so much as warring among themselves with the peasants taking the brunt as usual. And as if this was not enough, the countryside was filled with roving bandits.

Eventually, these external invasions were checked with the Norsemen settling down in Normandy and the Magyars in modern Hungary. Nevertheless, the destruction continued. Large landowners (read: warlords), seeking to cement alliances, rewarded their friends and retainers with bishop's robes and made them abbots of monasteries. The positions within the church brought with them rather good incomes, so at times, rather than give them as rewards, the warlord simply pocketed the money himself. The men who received these places generally had no interest in being a bishop or abbot, so they did not actually reside within their areas of responsibility. This gave rise to one of the great complaints of the later Reformers: absenteeism.

This continental disorder led to the marginalization of the papacy itself, which likewise sank into depravity, violence, and intrigue. Pope followed pope with startling rapidity as they were deposed, imprisoned, or murdered outright. Some popes were worse than the worst of the Roman emperors of old—and that's saying something. There were brief interludes attempting reform, but they were short-lived and soon forgotten.

A single example of papal depravity will suffice. A powerful family had Sergius III installed as pope in 904. He had beaten out two rivals, Leo V and Christopher I, both of which he promptly jailed and executed. The powerful family that had brought Sergius to fame was headed by one Theophylact and his wife, Theodora. These fine people had a daughter name Marozia who was Pope Sergius's mistress all the while married to the apparently permissive Guido of Tuscia. When Sergius died; John XII was made pope; but he was, after being incarcerated, suffocated with a pillow by Marozia and Guido. Following the brief pontificates of Leo VI and Stephen VII, Marozia caused to be placed on the papal throne her offspring from her liaison with the illustrious Sergius III who took the name of John XI. This sort of thing was not unusual. It went on for decades. You can't make this stuff up. Christ no longer lived in Rome. He had moved to the country.

The moral corruption of the papacy drew loud protests from round about as many were still true to the faith, but these shouts fell on deaf ears. In 991 a synod met and its chief Arnold of Orleans had this to say:

> Looking at the actual state of the papacy, what do we behold? John (XII) called Octavian, wallowing in the sty of filthy concupiscence, conspiring against the sovereign who he had himself recently crowned; then Leo (VIII) the neophyte, chased from the city by this Octavian; and that monster himself, after the commission of many murders and cruelties, dying by the hand of an assassin. Next, we see the deacon Benedict, though freely elected by the Romans, carried away captive into the wilds of Germany by the new Caesar (Otho I) and his pope Leo. Then a second Caesar (Otho II), greater in arts and arms than the first, succeeds; and in his absence, Boniface, a very monster of iniquity, reeking with the blood of his predecessor, mounts the throne of Peter. True, he is expelled and condemned; but only to return again, and redden his hands with the

blood of the holy bishop John (XIV). Are there, indeed, any bold enough to maintain that the priests of the Lord over all the world are to take their law from monsters of guilt like these men branded with ignominy, illiterate men, and ignorant alike of things human and divine? If, holy fathers, we be bound to weigh in the balance the lives, the morals and the attainments of the meanest candidate for the sacerdotal office, how much more ought we to look to the fitness of him who aspires to be the lord and master of all priests! Yet how would it fare with us, if it should happen that the man the most deficient in all these virtues, and so abject as not to be worthy of the lowest place among the priesthood, should be chosen to fill the highest place of all. What would you say of such a one, when you behold him sitting upon the throne glittering in gold and purple? Must he not be the "Antichrist," sitting in the temple of God, and showing himself as God? Verily such a one lacketh both wisdom and charity; he standeth in the temple as an image, as an idol, from which as from dead marble you would seek counsel.

Long years of ignorance, strife, and intrigue would stalk the official hierarchy of the church. Society in general would be just as chaotic and self-interested. The Crusades would be fought to regain the Holy Land from Islam as well as to get marauding bands of dispossessed soldiers the heck out of Europe. The church would later call for Crusades against other Christians. Somewhere along the way, maybe with Boniface VIII in 1302, the role of the apostles and their teaching in scripture were completely smothered. The decree of the pope and the taking of the sacraments superseded scripture in importance. In fact, scripture came to mean anything that the pope said that it did. The pope had become infallible, and he was the master of the faith. This was not a difficult position to maintain because village priests and their flocks were equally ignorant and unable to read the scriptures for themselves even had they been available.

During this time, the popes had managed to become huge landowners in their own right as did many bishops and abbots of monasteries. The clergy had become an attractive career choice. This is not to say that there were no leaders who held fast to the faith, but they were becoming fewer in number and their position was increasingly more difficult to maintain. But some of these became voices that were beginning to make themselves heard calling for reform.

By the fourteenth and fifteenth centuries, the church was in a sad state if you consider the glorious message of good news that it had been founded upon fifteen hundred years earlier. You remember the church, that entity established by the apostles and watered by the blood and tears of thousands of martyrs? The church had become rife with simony (the buying and selling of ecclesiastical offices), pluralism (the holding of multiple church offices by one man), and absenteeism (clergy not residing in their area of responsibility). In addition, *indulgences* were sold that the pope claimed could buy time off from purgatory or provide forgiveness of sins for both the living and the dead. The monies collected went to fill church coffers and/ or to finance papal wars. Enterprising entrepreneurs signed on as agents for the sale of these documents—receiving a cut, of course.

Wycliff

In the first quarter of the fourteenth century, a boy named John Wycliff was born in Yorkshire, England. Not much is known of his early years, but when he was about fifteen years old, he went off to study at Oxford where he became known for being a persistent user of logic. He became a well-respected scholar and eventually penned two books, which were widely circulated among the educated. The titles of these works were *On Divine Dominion* and *On Civil Dominion*. In *Divine Dominion,* he caused joy to abound among the temporal authorities by stating that the dominion of the church came from Christ and that those on whom it had been granted on earth were to use it to serve the people rather than being served by them. The example the clergy was to exhibit was to be that of a reflection of Christ himself. Taxes that were collected merely to enrich the church were, therefore, illegitimate. The magistrates crowed as they had long argued with Rome about the levying of such taxes. In fact, Wycliff served as a representative for the state at a conference held in Bruges in 1374 with members of the clergy in an effort to settle these disputes. The crowing by the magistrates abruptly ended when

Wycliff pronounced the same limits on civil authorities that he had prescribed for the church.

Perhaps the greatest service that Wycliff performed was his beginning the translation of the Latin Vulgate Bible into English. While he was fully convinced that the church was the guardian of scripture and had a more perfect ability to interpret it, he also believed that all brothers and sisters in Christ had the right of access. Scripture had long been held by the church to be its sole possession, and if they allowed the genie to escape the bottle, who knew what would happen?

Despite irritating both sides in this tax controversy, John Wycliff managed to balance his offenses fairly evenly so that he remained under the radar of major controversy with both church and state. This would change when he said that the true church of Christ was not represented by popes and bishops but by the invisible body of Christ physically represented by his redeemed people. This set some ecclesiastic teeth on edge but nothing like his teaching on communion. The doctrine of transubstantiation had been confirmed by the church in 1215 at the Fourth Lateran Council. This stated that the moment the bread was consumed it turned into the literal flesh of Christ. (Pretty sick, I think.) They believed the same with the wine, but only priests received the cup. The little people were not invited to partake. Wycliff taught that the body of Christ was, in fact, present during the taking of the sacrament but in a mysterious way. The bread was not replaced by flesh. It remained bread. As a result, he became numbered among the heretics by many of his colleagues but was not officially labeled as such by the church—yet.

Wycliff was caused further difficulties when the peasants revolted under Wat Tyler in 1381. Since he had sympathized with some of their grievances, many came to consider him the fomenter of the rebellion (which was not true). Finally, in 1382, some ten of his topics of teaching were officially declared heretical and all his writings were banned. While still widely admired, many of his devotees at Oxford and elsewhere deserted him. Wycliff was not officially excommunicated and died of a stroke two years later. As a sort of posthumous revenge, at the Council of Constance some churchmen were apparently embittered by his not being expelled from the church while he lived, so they had him roundly condemned in his grave, disinterred his body, and had it burned. His ashes were flung into the Swift River. This may have made these individuals feel better; but the genie was, in fact, out of the bottle. The first shots of the Protestant Reformation had been fired.

Hus

In the latter part of the fourteenth century, there existed strong ties between the kingdom of Bohemia (in the current Czech Republic) and England. These ties were the result of several circumstances. There was a pope in France and one in Rome (the Great Western Schism), and King Wenceslaus of Bohemia had recently been deposed as emperor with the agreement of the pope in Rome. Perhaps he was further annoyed by the fact that the new emperor was his half brother Sigismund. In any event, to vent his pique I suppose, he threw his weight into the balance on the side of the French pope.

Another circumstance that caused these close ties was the fact that Richard II of England was married to a Bohemian princess. Additionally, the faculty and students of the University of Prague were in close communication with those at Oxford. In those days, there wasn't an institution of "higher learning" in every city and town as there seems to be today. The communication back and forth between the two universities caused the writings of Wycliff to be widely circulated and ardently debated. The groups who debated these issues usually drew their lines between the Bohemians and the Germans probably because of a latent Czech nationalism and the tiresome influence of Germans in general. Eventually, the German students picked up their toys and went home to found their own university in Leipzig.

By this time, John Hus had become dean of the faculty of philosophy at Prague. While the previously mentioned debates were at their height, Hus defended the right of the students to read and discuss Wycliff but personally rejected his ideas around the sacrament of communion. He subscribed to the official doctrine of transubstantiation. His actual inclination at this point was to help restore the church to the original intent of Christian living with special emphasis on reforming the clergy.

Hus preached from the pulpit of the Chapel of Bethlehem, and he did so in the common language of the people. From the pulpit, he put his inclination into practice by railing against the enrichment of the church at the people's expense and the usual foibles of the clergy, which included fornication along with the time-honored simony and absenteeism. His attacks reached beyond the mere parish priest.

The higher echelons of the church felt that the teachings of Wycliff, which by this time had been banned, were being expounded upon by Hus. Preaching from the Chapel of Bethlehem was ordered to stop. Hus disobeyed and continued preaching while the church bosses ordered the books he had written to be burned. In 1410, he was ordered to report at Rome, which he also disobeyed, and the church countered the following year by excommunicating him. Hus became more radical in his views and stated that any pope that is unworthy, need not be obeyed. The idea of whether a pope was worthy or not centered on whether such a pope was acting in a self-interested manner rather than in the general interest of the church. Scripture was the final arbiter in judging such things, and all needed to live by its teaching, including popes.

In 1415, the emperor Sigismund invited Hus to the Council of Constance (the same council that had caused the body of Wycliff to be disinterred and burned) to defend his views. It was hoped that this council would be able to finally end the Great Western Schism and elect one pope to preside over the church from Rome. Sigismund guaranteed Hus safe conduct to and from the council. Having agreed to these stipulations, Hus appeared before the council on June 5, 1415. He was promptly called on to recant his published views and teachings. Considering that if he did so, he would not only admit to being a heretic, which he wouldn't do, but would also confer the same standing on his friends and followers, which his conscience would not allow. Regardless of his promise of safe conduct, the emperor stood by and saw Hus burned at the stake and his ashes thrown into Lake Constance.

Back in Bohemia, the people were in an uproar. Four hundred and fifty-two nobles gathered in an assembly of their own and publicly stated their agreement with Hus and refuted the Council of Constance. In retaliation, the council, which would be in session until 1418, ordered the university at Prague to close its doors and called the nobles to appear before them. Their numbers, despite the council's condemnation, continued to swell and several smaller groups joined them. With open fighting looming on the horizon, the Hussites (as they had come to be called) wrote and agreed to Four Articles:

1. The Word of God was to be preached freely throughout the kingdom.
2. The laity would receive both the bread and the cup during communion.
3. The clergy would rid themselves of their riches and live in apostolic poverty.
4. Gross and public sin would be punished and this included simony.

In 1419, Wenceslaus died, and Sigismund prepared to add king of Bohemia to his résumé. The Hussites insisted that he accept the Four Articles before they would allow him to assume the throne. Sigismund refused, whereupon the pope sent five—count them five—Crusades between 1419 and 1431 to stamp out the renegades. Each and every one on them was defeated. In order to negotiate with the victorious Hussites, a new council was called at Basel. The council, realizing the error of burning Hus (supposedly), invited his followers to come and meet with representatives of the church. Not to be duped, they demanded guarantees that the council considered to be of an offensive nature, so they sent off another Crusade, which was, yet again, defeated.

This time, the church negotiated in good faith and the church in Bohemia rejoined the rest of the flock with some concessions appertaining to themselves. Specifically, the laity could continue receiving the bread and cup and some other portions of the Four Articles were honored. Sigismund was finally able to ascend his throne, though he died shortly after, because many Hussites accepted these terms. Some, however, refused and formed their own church, the Union of Brethren (Unitas Fratrum), who would grow and add adherents over time and would eventually include members in Moravia. These folks managed to hang on through many trials and are called, simply enough, Moravians.

Wycliff and Hus were not the only voices of dissent in the fifteenth century, but they were the men who had largely set in motion the mission to rescue the gospel.

The Walls Come Down

"So, the people shouted when the priests blew the trumpets. And it happened when the people heard the sound of the trumpet, and the people shouted with a great shout, that the wall fell down flat" (Josh. 6:20a).

There were many things that caused the Protestant Reformation, not the least of which, I think, was the hand of God. Church leadership was largely out of control, and men like Wycliff and Hus had spoken out only to be martyred. In addition to the everyday practices of the Roman church, inquisitions and Crusades had also soured the people, not against the faith but those who had been charged with the care and propagation of it. When Martin Luther appeared on the scene, he did not free the church of Christ from legalism—Christ himself had already done that—but he reminded them of that freedom and the direct connection to God they enjoyed without the need for priests and popes. The church had lost the sense of this reality just as the Jews had lost the heart behind the Law by the time of Christ. It had become a religion of man, and the love of God, gratitude for the work of his Son, was only trotted out to placate the masses. Jesus, apparently, had been lost in the jockeying of the church for position and power.

Luther was a complex man, and to cover his life in these pages is not possible. We must take time, however, to briefly describe who he was and what he did because it is through him and his followers, largely, that Protestantism would come to represented a complete renunciation of popery.

Martin was born in 1483 at Eisleben into a strict home. He was raised not as a child of the Renaissance but in the most religiously conservative portion of the population. His family was of the working class, and his father spent long hours in the copper mines until he was able to purchase half a dozen foundries. Don't think that this family business brought them to the point of affluence because we are told that the wife had to daily go into the forest to drag home firewood for the family hearth. The Luthers were as rough as they were devout.

The University at Erfurt, where he sought to take his degree, was established along lines that mirrored his upbringing. Here they provided no "liberal arts" degree as we understand the term today. All of Luther's training, home, school, and university was centered on making him fear God and to reverence the church. His university contemporaries had been raised in much the same way. There was nothing that jumped out about Luther that would lead one to expect him to be markedly different from his peers. He received his BA in 1503 and his MA in 1505.

Luther

At an early age, Martin developed a tendency to be anxious and fall into depths of depression. The nature of God as portrayed in medieval times probably exacerbated these tendencies and made him extremely conscious of the condition of his soul. In fact, he suffered torments over it. God was seen, on the one hand, as a benevolent Father (a vision not easily recognized by Luther) and on the other as the wrathful consignor of mortal flesh to the pit of eternal wailing, flame, and torment. This irresistible Father could only be placated by his Son or softened by the Son's mother, Mary. Should Mary be indisposed, you could seek the assistance of her mother, Saint Anne. If all three of them were unwilling to take up your case, you still had recourse to a whole truckload of saints.

As stated previously, the church sold indulgences, which had caused Wycliff, Hus, and many others heartburn for some years. But the church also taught that Jesus had built up a huge bank account of grace by the time he ascended to heaven. This account was accessible to all worthy supplicants provided they performed certain tasks. One was the visiting of various locations, which were the repositories of holy relics. It seems like every church in existence had a piece of John's thigh bone or Peter's clavicle or a piece of the true cross or at least the dirt it had been set in. If one were to make a pilgrimage to one of the churches possessing such wonders, to see the piece of bone or wood or dirt, they could make a withdrawal from Jesus's account and relieve themselves or a departed loved one from some time in purgatory. Another choice attraction was the Scala Sancta in Rome.

The Scala Sancta (known at that time as the Scala Pilati [Stairs of Pilate]) are twenty-eight white marble steps that are reputed to be the stairs leading to the praetorian of Pontius Pilate in Jerusalem. Around 326, Constantine sent his mother, Helena, to the Holy Land in a search for relics. Thanks to Helena's diligence, it was reported that she had discovered three crosses, one of which was *the* cross. We are told that it was the true cross because it was subjected to scientific testing though you will be surprised by who did the testing and the test that was applied. Helena had a woman who was near death brought to the discovery sight and touch all three crosses. Upon having contact with the third, she felt wonderfully recovered. Luckily, for our ardent relic hunter, the nails also happened to be lying about close by. Old Helena was also said to have lugged home soil from Golgotha to mix with the dirt in the Vatican Gardens. Apparently, she also acquired the tunic worn by Christ while in Jerusalem and had it shipped back to Trier.

But the staircase was by far the biggest, heaviest object to be brought back by this indomitable woman. It is located in a building near the Archbasilica of St. John Lateran today but was originally installed in the Lateran Palace. In any event, it was said that by climbing these stairs on your knees, one by one, the climber's past sins would be forgiven. Roman Catholics continue the pilgrimage to this day as they can earn nine years' relief from purgatory for each step climbed. In 327, our Helena returned home and placed her other treasures in her palace's private chapel where they can still be seen today. Who knows, perhaps they are authentic.

Luther's father had sent him to university with the view of his son becoming an attorney. It was considered a lucrative-enough occupation that young Martin would be able to take care of not only himself but also his parents in their old age. But on July 2, 1505, something went terribly wrong with the father's plans. Six months after earning his MA, Martin was walking along a dusty summer road toward a village named Sotternheim. As he got closer to the village, a thunderstorm blew up (of the type we in the South call a toad choker), which rumbled and crashed all about him. A bolt of lightning landed so close to him that it knocked him to the ground. In sheer terror, he cried out, "Saint Anne, help me! I will become a monk!"

You see, Luther was a faithful son of the Catholic church. He believed in the efficacy of saints, popes, monasticism, and the observance of doctrine as dictated by the pope. It was, I think, the most profound irony that would make him repudiate nearly all of it as the years rolled on and help to firmly establish Protestantism. The pope, whom he now revered, would come to be seen by him as Antichrist.

True to his vow and against the vehement objections of his father, Martin showed up at the door of the Augustinian Monastery at Erfurt barely fifteen days after his experience with electricity. After two years of serving as a novice, he took the vows and put on the cowl. Luther entered a monastic order because he believed that here, if anywhere, he could make his peace with God. The daily routine of the monks was such that after sleeping eight hours they were awakened between one and two in the morning by the monasteries bell. A second intonation would summon them to church where they all entered the choir stall and chanted matins for three quarters of an hour. Their day was broken up into seven sections, and each began with prayer and ended with the cantor (leader of liturgical music) chanting:

> Save, O Queen, Thou Mother of mercy, our life, our delight and our hope. To Thee we exiled sons of Eve lift up our cry. To Thee we sigh as we languish in this vale of tears. Be Thou our advocate. Sweet Virgin Mary pray for us, Thou Holy Mother of God.

In this environment—completely resigned from the troubles of the world, spending their time in prayer and fasting, living an ordered life free of chaos—it is easy to see why Martin imagined himself following the footsteps of the saints who came before him . . . at least for a while.

The day arrived when Brother Martin was to conduct his first mass. He desperately wanted his father to attend as they had not seen each other since he announced his intention to become a monk. Hans Luther had apparently gotten over his disappointment because he arrived as a proud father bringing with him an escort of twenty horsemen and a generous contribution for the monastery. Martin took his place before the altar and began the introduction of the mass until he came to the words: "We offer unto thee, the living, the true, the eternal God." The emotion that nearly overcame him he relates himself:

> At these words I was utterly stupefied and terror-stricken. I thought to myself, "With what tongue shall I address myself to such Majesty, seeing that all men ought to tremble in the presence of even an earthly prince? Who am I, that I should lift up mine eyes or raise my hands to the divine Majesty? The angels surround him. At his nod the earth trembles. And shall I, a miserable little pygmy, say 'I want this, I ask for that'? For I am dust and ashes and full of sin and I am speaking to the living, eternal and the true God."

Another lightning bolt threatened to do him in. It was only by utilizing stern self-control that he could finish the mass. Luther recognized his unworthiness. He needed relationship with God to find his peace, but how could a pygmy stand before Holy God? Brother Martin was confronted with all the doubt, despair, panic, and desperation, which can come over a person whose status of soul is unknown and who deeply cares about it. Luther recommitted himself to the order. He would outpray, outfast, outmonk the other monks. Any grace that could be achieved via good works would be his. At one point, he sat at confession for six hours fearful lest he overlook something. An additional agony was added; what if he sinned and didn't recognize it as such? It would go unconfessed! He became more and more disillusioned.

Luther decided that he might access some of that bank account of Jesus, which was overflowing with grace for himself. In 1510, two brothers from Erfurt went to Rome to settle a dispute that had arisen among the Augustinians, the pope having been called upon to resolve it. Brother Martin was one of them. Roman history and antiquities held no interest for him, so upon his arrival, he immediately sought to take advantage of such attractions as might assist to save his soul.

While he had some duties at the cloister where he was staying, they were not time-consuming. He ran hither and yon to hit all the sacred catacombs, basilicas, bones, and every other thing that offered grace by visiting it, touching it, or praying to it. His intent being to amass as much grace as possible so that he could carry it home and dole it out between himself and his family.

A man with this fierce a concern for his soul and the soul of others was destined for shipwreck when confronted with the reality of Rome. Those priests who heard his confession were incompetent and disinterested. Especially the native Italian priests were viewed as frivolous and completely ignorant, running through six or seven masses while Luther was completing one. The priests, you see, were committed to a certain number of masses per day—not for the living worshipers but the dead. In addition, it was obvious that certain priests were blatant unbelievers and Brother Martin could not have failed to hear of the immorality of the Roman clergy. Rome, of course, had its district of ill repute like any city, but the horror of it came to Luther when he found that the ecclesiastics were frequenters.

It was while climbing the Sancta Scala on his knees, one step at a time, and repeating a Pater Noster and kissing each step that at top he cried, "Who knows whether it is so?" This was terribly disconcerting because if this did not do any good, who could tell if any of the other gyrations he had gone through did? Luther said that he had gone to Rome with onions and had returned with garlic.

A great change was waiting on Luther when he returned home; he was transferred to Wittenberg from Erfurt. Compared to Erfurt, Wittenberg was a backwater, at least in population, but the land itself was very productive. And Luther was to benefit greatly from the elector, Frederick the Wise, who was in love with his newly founded university and dreamed of its growing to rival the one in Leipzig. Frederick had invited the Franciscans and Augustinians in 1511 to provide him with three new professors. One of them was Luther. Here he met the man who would be a positive source of encouragement and advice: Johann von Staupitz.

Sometime around his appointment to Wittenberg, Martin's anguish reached new heights and left him at times subject to tremors. The timing of them and the exact nature of their cause remain unknown, but what is known is that he used every notion or doctrine of the church to try to soothe a soul consciously tormented by an alienation from God. Trying good works, he found that he could never do enough of them. He prayed more often, confessed more often, and performed penance more often; but all were unable to remove the doubts which deprived him of any assurance of salvation. Luther had reached an impasse. In his anguish, he cried, "Is it not against all natural reason that God out of his mere whim deserts men, hardens them, damns them, as if he delighted in sins and in such torments of the wretched for eternity, he who is said to be of such mercy and goodness? This appears iniquitous, cruel, and intolerable in God, by which very many have been offended in all ages. And who would not be? I was myself more than once driven to the very abyss of despair so that I wished I had never been created. Love God? I hated him!"

Martin finally sought to understand the fullness of scripture. In 1513 he started to lecture on Psalms followed in 1515 by Romans and then Galatians in 1516–17. In the Twenty-Second Psalm, presaging the Passion of Christ, verse 1 states: "My God, My God, why have you forsaken me? Why are You so far from helping Me, and from the words of My groaning?" Jesus, on the cross, had experienced the same despair as Luther. How could this be explained? It could only be that Christ had taken on Himself the sin of us all! This was an entirely new picture of Christ to Luther. The mighty judge and fearsome, wrathful God was also the broken Man on the cross. The debt of us all had been paid by the very God that would otherwise condemn us.

How could this be understood? Man's wisdom had indeed been made foolish. Reason must be set aside and replaced by faith. This gospel—long hidden under vestments, superstition, and ritual—was marvelous to behold. It was a revelation. God must first allow us to suffer so that he might turn us toward himself for the alleviation of it. Martin came to accept that the process of becoming reborn was, as Paul said, "justification by faith." Luther said:

> I greatly longed to understand Paul's Epistle to the Romans and nothing stood in the way but that one expression, "the justice of God," because I took it to mean that justice whereby God is just and deals justly in punishing the unjust. My situation was that, although an impeccable monk, I

> stood before God as a sinner troubled in conscience, and I had no confidence that my merit would assuage him. Therefore, I did not love a just and angry God, but rather hated and murmured against him. Yet I clung to the dear Paul and had a great yearning to know what he meant.
>
> Night and day, I pondered until I saw the connection between the justice of God and the statement that 'the just shall live by his faith.' Then I grasped that the justice of God is that righteousness by which through grace and sheer mercy God justifies us through faith. Thereupon I felt myself to be reborn and to have gone through open doors into paradise. The whole of Scripture took on a new meaning, and whereas before the 'justice of God' had filled me with hate, not it became to me inexpressibly sweet in greater love. This passage of Paul became to me a gate to heaven.
>
> If you have a true faith that Christ is your Savior, that at once you have a gracious God, for faith leads you in and opens up God's heart and will, that you should see pure grace and overflowing love. This is to behold God in faith that you should look upon his fatherly, friendly heart, in which there is no anger or ungraciousness. He who sees God as angry does not see him rightly but looks only on a curtain, as if a dark cloud had been drawn across his face.

Luther's initial focus of reform was teaching the Bible, and this he proceeded to do among his peers in Wittenberg and in the chapels in which he was called to conduct services. His theology would be that of Paul forever after, made sharper and provided with clarity to those he sought to teach. This is not to say that he was unaware of the disparity between scripture and what he witnessed in Rome and elsewhere in the church. Contained in his lectures were harsh criticisms of the greed and sinfulness of churchmen in high places, with particular emphasis on then-pope Julius II.

Quietly and persuasively, Luther turned most of his colleagues in Wittenberg toward the understanding of scripture as the final authority on any matter theological, above pope, or man-devised doctrine. In order to conduct an orderly debate, he penned ninety-seven theses to be discussed in an academic environment, and since Latin was the language of academic debate, he wrote them in that language. He was disappointed to discover that they were attended with but little interest. He wrote another set, this time of ninety-five theses, also in Latin, and was now conditioned to expect a similar response. These ninety-five theses, however, were discovered by an unknown person, translated into German and widely circulated.

The document was entitled *The Ninety-Five Theses on the Power and Efficacy of Indulgences.* Contained within he attacked not only indulgences as being worthless but also anybody having anything to do with the peddling of them, including the temporal authorities. In addition, he did not fail to mention that their chief value came in the way of profit to those who sold them. Of course, both the pope and the emperor were enraged, perhaps with different reasons but enraged nonetheless, and sought to have him silenced.

The pope wanted the Augustinian order, to which Martin still belonged, to handle this rather prickly situation. The next Augustinian assembly was held in Heidelberg to which he was accordingly summoned. Luther fully expected to be tried as a heretic and consigned to the flames shortly thereafter, if not sooner. Much to his surprise, contrary to his gloomy expectations, he found that many of the brothers favored his view of scripture and the younger ones accepted it with enthusiasm.

This tactic having failed, the pope, Leo X, decided to have Luther summoned to an imperial diet to be presided over by the emperor Maximilian who had also been deeply offended by the reformers comments. Once again, Luther walked into what he thought was to be his condemnation and extermination. At the diet, he was not permitted to defend his views but, like John Hus before him, ordered to recant. Again, like Hus, he said he would be happy to do so if someone would convince him in scripture that he erred. His accuser, a representative of the pope named Cajetan, was not open to debate and had papal authority in hand to arrest Luther. When Martin became aware of this warrant, the Wittenberg teacher slipped away at night and returned home and to his protector, Frederick the Wise.

Finally, after much to do, a papal bull was issued, ordering the wayward monk to either recant or be excommunicated. The response was deliberate and dramatic. He publicly burned the bull along with some of the most offending books representative of "popish doctrines." The break with Rome was absolute and permanent. The only question remaining was whether Luther and his teaching would survive. He was thereupon called before a new assembly at Worms, which would be overseen by a new emperor as Maximillian had died. This new monarch was Charles V who was Spanish, a staunch son of Rome and possessed of a none too high an opinion of his German subjects. Brought before this assemblage of high-profile gentlemen, he was unceremoniously ordered to recant. His response was "My conscience is a prisoner of God's Word. I cannot and will not recant, for to disobey one's conscience is neither just nor safe. God help me. Amen."

Offending a pope was one thing, but Charles could not abide the idea of being thus spoken to by a ragged monk. He resolved to do something regardless of the safe conduct that been promised to Luther upon his appearing. (We know the value of such promises by the same having been "guaranteed" Hus.) The following edict was issued: "Luther is now to be seen as a convicted heretic. He has twenty-one days from the fifteenth of April. After that time, no one should give him shelter. His followers also are to be condemned, and his books will be erased from human memory."

Unbeknownst to Charles, Frederick had learned of the intent of the men at Worms and devised a plan of rescue for the monk/ professor. An armed band abducted Luther and secreted him in Wartburg Castle. Frederick himself did not know where he had been taken—plausible deniability. It was rumored that the pope had ordered him killed, some said the emperor. Apparently, Frederick cared more for his word than did Charles for it was through the elector that the safe conduct been granted. Perhaps Frederick did not care for the Spaniard Charles any more than Charles did the Germans. In any event, Luther was safe.

While he was tucked away at Wartburg, he spent his time writing. What was arguably his most significant achievement, he translated the Bible into German. The New Testament was accomplished in two years while the Old would take him ten. His countrymen would appreciate the effect long after the writing because he did much to consolidate the many dialects by carefully choosing his words. Germans now had a Bible of their own and in the vernacular.

Meantime, much was occurring in the German domains of Charles V. While he had every intention of putting down the religious dissent in these lands, something always seemed to come up to thwart him. For one thing, Leo X had died, so Charles finagled his boyhood tutor into the position of pope. His papal name was Adrian VI, and he would not accept any deviation from the traditions of the Roman church. Adrian was very interested in reforming the behaviors of churchmen, however, and he actively sought to instill a program of austerity. Unfortunately, Adrian only lasted a year and a half and was succeeded by Clement VII who entertained no similar interest in a curtailed lifestyle.

When next Charles sought to stamp out the Lutheran "heresy," Suleiman was marching on Vienna, so he thereupon set aside religion for the moment to gain the aid of his German subjects in his war against the Turks. After dealing with this crisis, Charles returned and once again had the Lutherans on his radar, but in his absence (1522–23), another situation had occurred. The German knights had become rebellious because of their declining fortunes, so they were beginning to coalesce around a growing sense of nationhood. They had a leader, one Franz von Sickingen, and were apt to blame Rome for their landlessness and poverty and saw Luther as representing a new Germany. Many claimed that they were fighting for the Reformation when they finally rebelled, but the reformer had nothing to do with encouraging their actions.

Not to be outdone by the knights, in 1524 the peasants also rebelled. This was nothing terribly unusual as sporadic rebellions had been seen before only on a more localized level. This time it was much more widespread and became particularly nasty when they took it into their heads to say that the reformers felt them justified. This uprising resulted in a drafting of *Twelve Articles*, which sought to base their demands on scriptural grounds. Martin was caught in the middle of this as the peasants sought his agreement with their claims. When he read the *Twelve Articles*, he told the princes that he thought them just, but he personally had a horror of equating scripture with armed insurrection. Many of the peasants, as a result, felt that he had betrayed them and returned to the church of Rome or became Anabaptists. (Anabaptist advocated baptism and church membership for adults only, a refusal of all oaths, and nonresistance or pacifism.)

The upheaval of the Protestant Reformation saw monks, priests, and nuns give up their vows and not only leave the church of Rome but also marry and become fully engaged in secular life. Some took to the new churches as pastors where such positions were available (and they were wanted), and others assumed trades or farming. As he felt he must lead, Luther too took a wife. Her name was Katharina von Bora (who was herself an escaped nun), and together they had six children. Initially, he was reluctant to take a wife as he still expected to become a martyr at any time. Their union became one that was quite full and happy. Martin Luther continued to write, teach, and raise his family alongside Kate. Luther died in 1546, but his legacy is carried on in the Lutheran traditions today.

John Calvin had an equally prominent role in the Reformation, and his teaching has had an even greater impact on the generations that followed. Born in Noyon, Picardy, France, in 1509 he was a young contemporary of Luther's. Calvin broke away from the Roman Catholic Church in 1530 after Protestant Christians in France were brought under persecution. Facing some unpleasantness of his own, he fled to Basel in Switzerland, and it was here in 1536 that he published the first edition of the *Institutes of the Christian Religion*. In this same year, he was invited by another Frenchman, William Farel, to help bring the reformed church to Geneva. The city council resisted the new teaching and expelled both Calvin and Farel. Calvin turned to Strasbourg to minister to a church composed of French refugees but maintained his contact with the church in Geneva. In 1542 he was invited back to that city, and there he remained until his death in 1564.

Calvin

Needless to say, it is the theology of Calvin that is of most interest to us. His writings were voluminous and neatly articulated and well worth the reading, but we will focus our attentions on what has been passed down to us as the Five Points of Calvinism. These five points have also been remembered by the acrostic TULIP, and we will investigate them using this method.

Total depravity: There is no part of man that is not infected with sin. We are completely sinful. Our hearts, bodies, minds, and emotions are all corrupted. We are not as sinful as we could be, but we are nevertheless completely affected by sin. Calvin based his conclusions, as he always did, on scripture.

> Mk. 7:21–23: "For from within, out of the heart of men, proceed evil thoughts, adulteries, fornications, murders, thefts, covetousness, wickedness, deceit, lasciviousness, an evil eye, blasphemy, pride, foolishness: all these evil things come from within, and defile the man."

> Rom. 6:20 : "For when ye were the servants of sin, ye were free from righteousness."

> Rom. 3:10–12: "As it is written, There is none righteous, no, not one: There is none that understandeth, there is none that seeketh after God. They are all gone out of the way, they are together become unprofitable; there is none that doeth good, no not one. Their throat is an open sepulchre; with their tongues they have used deceit; the poison of asps is under their lips."

> 1 Cor. 2:14: "But the natural man receiveth no the things of the Spirit of God: for they are foolishness unto him: neither can he know them, because they are spiritually discerned."

> Eph. 2:15: "Having abolished in his flesh the enmity, even the law of commandments contained in ordinances; for to make in himself of twain one new man, so making peace."

> Eph. 2:3: "Among whom also we all had our conversation in times past in the lusts of our flesh, fulfilling the desires of the flesh and of the mind; and were by nature the children of wrath, even as others."

In light of all scripture has to say about our natures, Calvin asked, "How is it possible that any should choose or desire Christ?" The answer is that we won't. God must handle this as well which he does through predestination. (It is probably the concepts of predestination and election that are the most consistently debated points of Calvinism. We will not take up that subject here as it will not advance our purpose.)

Unconditional election: There is nothing in any one of us that warrants our being selected above anyone else. He chooses solely on the basis of his kindness and sovereign consideration. It is not based on any merit we possess. God does not have to look into the future to see who will select him because, left to ourselves, none of us would. Also, as some of us are chosen, it is certain that some are not.

> Eph. 1:4–8: "According as he has chosen us in him before the foundation of the world, that we should be holy and without blame before him in love: Having predestinated us unto the adoption of children by Jesus Christ to himself, according to the good pleasure of his will, to the praise of the glory of his grace, wherein he hath made us accepted in the beloved. In whom we have redemption through his blood, the forgiveness of sins, according to the riches of his grace; wherein he hath abounded toward us in all wisdom and prudence."

> Rom. 9:11: "For the children being not yet born, neither having done any good or evil, that the purpose of God according to election might stand, not of works, but of him that calleth."

> Rom. 9:15, 21: "For he saith to Moses, I will have mercy on whom I will have mercy, and I will have compassion on whom I will have compassion. Hath not the potter power over the clay, of the same lump to make one vessel unto honor, and another unto dishonor?"

Limited atonement: Jesus's sacrifice was certainly sufficient for all mankind, but it was not efficacious for all. In other words, he died only for those who had been elected. The following verses speak to Jesus having died for "many," Jesus died for the sheep, Jesus prayed for those that are given to him, and the church was purchased by Christ.

> Matt. 26:28: "For this is my blood of the new testament, which is shed for many for the remission of sins."
>
> Jn. 10:11, 15: "I am the good shepherd: the good shepherd giveth his life for the sheep. As the Father knoweth me, even so know I the Father: and I lay down my life for the sheep."
>
> Jn. 17:9: "I pray for them: I pray not for the world, but for them that thou hadst given me; for they are mine."
>
> Acts 20:28: "Take heed therefore unto yourselves, and to all the flock, over the which the Holy Ghost hath made you overseers, to feed the church of God, which he hath purchased with his own blood."
>
> Eph. 5:25–27: "Husbands, love your wives, even as Christ also loved the church, and gave himself for it; that he might sanctify and cleanse it with the washing of water by the word, that he might present it to himself a glorious church, not having spot, or wrinkle, or any such thing; but that it should be holy and without blemish."

Irresistible grace: Those who are elect cannot refuse the call. God offers the truth of the gospel to everyone as an external call and which may be, and is, resisted. But to the elect, the call is an internal one that compels acceptance. This inward call is one whereby the Holy Spirit works on the heart and mind of the individual to bring him/her to a point of repentance and regeneration; thus, they willingly and freely come to God.

> Rom. 9:16: "So then it is not of him that willeth, nor of him that runneth, but of God that showeth mercy."
>
> Phil. 2:12–13: "Wherefore, my beloved, as ye have always obeyed, not as in my presence only, but now much more in my absence, work out your own salvation with fear and trembling. For it is God which worketh in you both to will and to do of his good pleasure."

> Jn. 6:28–29: "Then said they unto him, What shall we do, that we might work the works of God? Jesus answered and said unto them, This is the work of God, that ye believe on him whom he hath sent."
>
> Acts 13:48: "And when the Gentiles heard this, they were glad, and glorified the word of the Lord: and as many as were ordained to eternal life believed."
>
> Jn. 1:12–13: "But as many as received him, to them he gave power to become the sons of God, even to them that believe on his name: which were born, not of blood, nor of the will of the flesh, nor of the will of man, but of God."
>
> Jn. 6:37: "All that the Father giveth me shall come to me; and him that cometh to me I will in no wise cast out."

Perseverance of the saints: Here, Calvin emphasizes the concept of once saved, always saved. Put another way, you cannot lose your salvation. The Father has elected you, the Son has redeemed you, and the Holy Spirit has imparted to you the promised salvation. With these vouching for you, nothing can remove the seal of your salvation; you are eternally secure in the arms of Christ. Calvin took this stance based upon, among others, the following verses:

> Jn. 10:27–28: "My sheep hear My voice, and I know them, and they follow Me. And I give them eternal life, and they shall never perish; neither shall anyone snatch them out of My hand."
>
> Jn. 6:47: "Most assuredly, I say to you, he who believes in Me has everlasting life."
>
> Rom. 8:1: "There is therefore now no condemnation to those who are in Christ Jesus, who do not walk according to the flesh, but according to the spirit."
>
> 1 Cor. 10:13: "No temptation has overtaken you except such as is common to man; but God is faithful, who will not allow you to be tempted beyond what you are able, but

> with the temptation will also make the way of escape, that you may be able to bear it."

> Phil. 1:6: "Being confident of this very thing, that He who has begun a good work in you will complete it until the day of Jesus Christ."

Calvinism had some detractors at least as far as agreeing with all five points. Chief among these is Arminianism based on the teaching of Jacobus Arminius a.k.a. Jakob Harmenszoon (1560–1609), a leader of the Dutch Reformation who studied under Theodore Beza, Calvin's successor. As previously noted, the most pronounced point of contention was that of predestination and election of the saints. It would be appropriate at this point to look at this controversy a little closer.

Predestination basically states that all events have been willed by God especially the disposition of individual souls. Many have made a rather a loud noise that this excludes us from making free choices; our wills are not free since God has already decided everything. But I am led to ask, what is that to us? Do *we* know what God has decreed beyond what is given to us in scripture? How do *we* know who God has elected and/or preordained? We don't. In *our* reality, we still need to choose. We do not know the mind of God, and neither can we. He doesn't let us in on what he has done. His reality is not our reality.

There was much turmoil along the way with various denominations rising. These denominations were largely regionalized because of the fact that while these events were taking place in Germany, similar reforms were going on in Geneva and other parts of Switzerland, France, England, and elsewhere. As they were contemporary with the Lutheran movement, they sometimes met and compared notes on their respective views of scripture. They did not agree on all the details, but one thing they were unanimous on was the gospel of Jesus Christ. Reunited to God through the mercy and grace extended to them through the dead, buried, and resurrected Christ. Faith in him first and last.

The Protestant Reformation has not ended. Today many of the old complaints have crept back into the church. Money, show and programs have taken the place of the Roman indulgences, ritual and infallibility in many of our modern churches.

Chapter Eleven

Q&A

I had promised earlier that I would address your questions. Well, you probably realize that I am unable to take questions directly, so what I really meant is that I would attempt to answer some commonly asked questions. Over the centuries, devout Christians have indeed been bothered by a number of issues, and personally, I don't find that simply being told to "have faith" is adequate. At least it wasn't for me. While I am not a trained theologian (which has its good and bad points), I will share with you what conclusions I have reached and how I arrived at them.

Unanswered Prayer

Many people have experienced their prayers simply not being answered and as a result have given up on them and, in some cases, God. Today, televangelists and promoters of the *prosperity gospel* tell folks to just pray harder and to send in their money and they'll receive "special prayer help" from that individual or organization. I'm not going to waste words on people who pray to win the lottery, big money on Wheel of Fortune, or the bass boat being raffled at

Bass Pro Shops. It's easy to become confused by all this nonsense. However, even Jesus said:

> And whatever you ask in My name, that I will do, that the Father may be glorified in the Son. Jn. 14:13
> That whatever you ask the Father in My name He may give you. Jn. 15:16b
> Most assuredly, I say to you, whatever you ask the Father in My name He will give you. Jn. 16:23b

Now, Jesus doesn't usually repeat himself unless he wants to stress a particular point. Apparently, he wished to stress the point of God's willingness to answer prayer and not just his willingness but his guarantee of an answer. But if you notice, this is not an unqualified promise. There are times when God *cannot* answer a prayer. (Folks who are still hung up on "God can do anything" are screaming. See chapter 8.) For example, imagine two men, equally qualified, who are applying to the same church for a vacant pastoral position. They are both vehemently praying for the position, but obviously, both cannot receive the job. One *must* be disappointed.

Much more common, however, is what happens on a day-to-day basis. All of us know of, or have been involved in, ardent prayer for someone who is seriously ill. There have been a couple of occasions I know of where the people who were leading the prayer chain had convinced themselves that a miracle would occur that would heal the individual being prayed over. The poor person died anyway.

How could Jesus make a promise like that about prayer and then not honor it? Maybe Jesus isn't really who and what he claimed. How do we reconcile—can we reconcile—the perceived disconnect between the prayer promise and our actual experience?

Many believers (myself included at one period) simply prefer to gloss over negative responses to prayer and say that God did answer; his answer was no. Others decry a lack of adequate faith for not getting their prayers responded to in a positive way. Still others merely claim that all prayers are answered positively whether we can see it or not. But these explanations simply will not do. At least these didn't satisfy me. When I first heard them, they sounded like "church speak" and not any form of explanation at all.

There must have been something in Jesus's comments that I just didn't pick up on. I had come to accept that Jesus simply would not intentionally confuse me let alone lie. As God, he *can't* lie. There must be some qualifications that are necessary to actualizing the fullness of his promise of answered prayer.

What might some of these qualifications be? First, it just makes sense, or so it seems to me, that if we are to be praying in the name of one holy (Jesus), we cannot be holding any sin in our lives which is unconfessed. Also, selfish motives are certainly not cool; and if we're honest, we may try to convince ourselves that our motives are selfless when they are not. But we need to dig really deep to determine whether any whiff of selfishness is involved. Speaking for myself, if I am particularly introspective, oftentimes at bottom, some selfishness lurks. It may not be, and most often isn't, intentional but it's there nonetheless. Also, timing can be an issue. We want a positive response *now* and are unwilling to faithfully wait on God's timing. If an answer that is so direct as to be obvious is not forthcoming within a day or two, we're in severe despondency. Let's face it, we have become a drive-thru society that tells time in milliseconds. At other times, I find myself not as earnest and persistent in desiring the outcome as I might be. (You see, I'm not pointing any fingers.) If I'm not earnest and persistent, God is aware that the outcome is not really all that important to me. Shame on me, I know it's true.

Finally—and this is the big enchilada—are we praying for, and in, the will of God? John tells us:

> Now this is the confidence that we have in Him, that if we ask anything according to His will, He hears us. And if we know that He hears us, whatever we ask, we know that we have the petitions that we have asked of Him. 1 Jn. 5:14, 15

Just what does it mean to be praying in the will of God? For one thing, are we praying for the advancement of Jesus's kingdom? Are we praying for the lost of this world? Are we praying for the glory of God to be revealed in all our activities, projects, relationships, and our very lives? These are the things God wants us to be concerned with. As we draw closer to him, as our desires become more closely aligned with his ends, we find that our priorities begin to change in a big way, and our prayer requests and the very tone of those prayers change as well.

My intent is not to try to tell you how to pray or what to pray for. Jesus has already done that in Matthew 6:9–13. What I am attempting here is to describe to you how I have seen things come around as I have come around. The purpose of God for his children is not for us to be continually happy. For sure we can allow ourselves the twenty-four-hour joy of knowing and belonging to him, but our comfort is not his primary motivation. Character building is more in his line to chisel and sand and smooth us until we more closely resemble Christ. That is what he wants us to want. Sometimes God lays aside the chisel and sandpaper and picks up the chainsaw. If so, we can be confident that it is for the good of those who trust him. He is not the great Santa Claus in the sky that checks our wish list on a regular basis. He wants us to return to the garden with our desires in line with what he wants for us. Christ did not die so that we might win the lottery.

We are told that God has a plan for us, and we are told that it is for good. I believe that many take that "good" to read instead "stuff." Have you ever stopped to consider that God's plan for us may be how we fit into *his* plan? Praying for simple personal things is not bad. It is not sinful. We are told to take all our concerns to the foot of the cross. But don't make these things the foundation of your faith or the getting of them the motivation to pray. Don't let getting what you want be conditional to your believing in the truth of Christ. Belief and/or faith ultimately is about knowing that whatever God does to, or for us, is better than anything we could want or, for that matter, imagine.

Suffering and Evil in the World

This is certainly not going to be adequately resolved in a few paragraphs, but I would be dodging the elephant in the room if I failed to mention it. Many people use this as the foundational reason that they cannot believe in God. How can an all-good, all-loving, merciful God allow all the hateful things that have transpired, and continue to transpire, in the world? Either he isn't there, or he lacks the power to do anything about it. We touched on this earlier. Even the church has perpetrated evil in the world as during the Crusades, the inquisitions, and the bloody attempts to prevent and then to crush the Protestant Reformation. And what of the churches turning not only a blind eye to Hitler but, as in the case of much of the church in Germany, siding with his regime? Shucks, you don't have to be a student of history to follow the trail of man's inhumanity to man. Just pick up a newspaper today. Well, pick up a newspaper that reports the news, that is. There are many forms of evil that clearly manifest the moral depravity of man.

There is another source of suffering that results from natural disasters. Tsunamis, floods, earthquakes, fires—hardly a week seems to go by that at least one of these isn't occurring somewhere. Regardless of the source, the question remains the same. How can God allow it?

Intellectually, we need to come to a rational understanding of this apparent dilemma. The atheist insists that if there were a God, he would be capable of creating a world wherein evil and suffering did not exist and would, in fact, prefer such a world. It does not necessarily follow that he could, in fact, create a world where everything was always good and still provide human beings with free will. If man had been created wherein he had no alternative but to choose rightly, he would not be free. As mentioned previously, I doubt the willingness of many folks to surrender their individual freedom of will to help a man unknown to them that lives in Siberia or on an island in the south Pacific. The atheist would find especially repugnant the loss of freedom to think and chose as they please.

Some atheists are willing to admit that the existence of God and evil simultaneously may not constitute a contradiction, but they complain that this God would surely limit the extent of evil and suffering. (Interesting that an atheist would pretend to know what God should and should not do.) But who is to say that the evil in the world would not be significantly worse if God weren't, in fact, exercising his power to limit it? In point of fact, there is no way to prove that God does not have sufficient moral grounds for the amount of suffering present in the world. But when all else fails, the skeptic merely claims that the existence of God, when viewed under the microscope of suffering and evil, is improbable.

Being unwilling to even let a claim of improbability rest unchallenged, a Christian has several facets of Christian doctrine that he can advance utilizing *all* we know about God. These would include the following:

1. Our chief purpose on earth is to seek out and to know God, not our personal happiness. God's job is not to keep us in perpetual mirth but to lead us to know the truth of him and to prepare us for eternity. In the same way, we as parents are not charged with keeping our children constantly entertained but to raise them to be responsible adults and—prayerfully—Christians.
2. Man is in a constant state of rebellion against God. We are fallen, narcissistic creatures. From birth, our focus is in attaining that which we want, when we want it, and how we want it. We are in need of correction and discipline. The wonder is that we do not receive all that we truly deserve.
3. God's purposes extend beyond our earthly existence and into eternity. Our lives can be viewed as boot camp for heaven. We are being put into shape, so to speak, to receive the incomparable riches of our eternal reward. One of the larger problems we have with this is our tendency to live in the moment. Few of us spend as much time contemplating the hereafter as we do playing golf, watching TV, spending time on Facebook, or playing video games.
4. The knowledge of God cannot be measured. Paul's (formerly Saul) personal experiences are particularly telling in this regard. He considered everything that didn't lead to Christ as worthless. In addition, he longed—yes, longed—to see his Savior face-to-face, and everything that kept him from that day was to be only tolerated and spent in the service of others.

Another point, not without significance, is that if everything was always beautiful, nobody would seek out God. The greater good, it can be argued, is more adequately served by the evil in the world that prompts us to look to God who is our source of strength and comfort. But the long story told short is that, in the end, we cannot know how God uses the fact of evil and suffering to fulfill his plan for mankind.

Consider that what may happen to you or me today may influence a grandchild or friend of that grandchild or someone who reads of the incident in a newspaper to come to Christ and lead others the same way. If you could know that ten, twenty, a hundred people would come to a reborn relationship with God because of some suffering on your part, would it not be worth it?

It is also worth noting that in questioning thus, we are asking God to justify himself to us. When we stop to consider the suffering of Christ and the unmitigated evil that befell him (we know it's unmitigated because we are told that God the Father did not temper it), how can we rail against God? Christ, the truly and wholly innocent, suffered because of *us*. Evil and suffering were dumped on him in ways that we cannot even guess at. The example of Christ, which we are to take as our own, is not one of ranting against all the evil and unfairness in the world but one of submission to the knowledge that all is according to the purposes of God. If you are a Christian, you of all people should be overjoyed at the fathomless good that came from the evil perpetrated on Christ.

How about Abortion?

Again, we touched on this before, but I think it deserves some further investigation. There are, I think, two things that we need to consider when determining how we approach this issue, and they are (1) do human beings possess an intrinsic moral value, and (2) is a human fetus in the womb a human being?

In answer to the first, while there have certainly been individuals, rulers, times, and places that perverted the concept of intrinsic human value, it has generally been recognized throughout recorded human history that we do possess such value. We'll let that question lie where it is on the merits of history.

Is the fetus a human being? Today, it is pretty common knowledge that at conception, all the information for a human being, if left to follow its normal course, is present. Individually, the sperm and egg each contain one-half of the necessary chromosomes and, if left to themselves, will die and be expelled. At the instant of conception, however, sex, facial features, hair and eye color, body type, etc., are established.

There are those who state that since the fetus has not left the birth canal, it is not a baby and therefore it is morally acceptable to destroy it. It is much less humanizing to refer to is as a fetus than a baby. By arguing in this manner, they completely remove the focus from the fact of the babies being human. Simply because one decides to focus on a particular stage of development of the child makes it no less a child. In conjunction with this, the argument for early-stage disposal is nothing less than a lie. Abortions are usually performed during the tenth to twelfth weeks when hands, fingers, feet, toes, veins, and movement are clearly distinguishable. In addition, the figure of 150,000 abortions annually is mentioned as taking place in the second and third trimesters (fourth to sixth month and seventh to ninth month respectively).

Of course, the battle cry of today's prochoice segment of society is a woman's right to choose what she does with her own body. (Does this include suicide? I wonder. I believe that's still illegal.) It is rightly pointed out, however, at no time is the fetus actually a *part* of the female body. The fetus is attached to the wall of the uterus and is a distinct living organism merely being nourished by its *host*. This argument seems to equate the removal of a human baby with that of a gallbladder or unsightly mole.

Others seek to define a person as a self-conscious individual. The fact is conveniently overlooked that this fetus will never have the opportunity to become that self-conscious individual because its development to that stage has been arrested by the mother, a doctor, and perhaps a nurse. And as ghastly to any right-minded person as it may seem, there are those who insist that it is right to kill it at the last instant insisting on a distinction between *being* a person and *functioning* as a person. These women, doctors, and presumably nurses don't mind sucking the brain out of the child before the head leaves the vaginal opening.

In fact, in 1973, a Nobel laureate of all things, James D. Watson, claimed that it would be morally acceptable to wait until three days after the birth to declare a child to be living! This would allow the parents time to decide if they wanted to keep it or to save themselves "a lot of misery and suffering." This, of course, is nothing short of infanticide.

Unfortunately, even Christians have been tainted with the evil, wholly worldly view that a woman has a right to choose. If we are Christians, we have surrendered whatever rights we may think we have to Christ. This point of view is unsustainable for anyone who claims the Bible as the inspired, holy, inerrant Word of God. Genesis 1:27 states, without equivocation, that man and woman are "created in the image of God." Genesis 9:6 says: "Whoever sheds man's blood, by man his blood be shed; for in the image of God He made man." And if you still wonder when God considers a child to be a human being:

> For you have formed my inward parts; you covered me in my mother's womb. I will praise You, for I am fearfully and wonderfully made; marvelous are your works, and that my soul knows perfectly well. My frame was not hidden from You, when I was made in secret, and skillfully wrought in the lowest parts of the earth. Your eyes saw my substance, being yet unformed. And in Your book they all were written, the days fashioned for me, when as yet there were none of them. Ps 139:13–16

Is Homosexuality All Right with God?

This has become a pervasive issue of the past twenty-five years or so. There are segments of denominations within the church who not only find tolerance for it but openly condone it. Some who say they are Christians view any biblical injunctions against this behavior as being either outdated or, they claim, absent from the text altogether. Of course, anyone who publicly states homosexuality as "wrong" is instantly branded a bigot and covered with the basest forms of ridicule. If you have stayed with me this far, you know that accusations should not sway us, but we should resort to what the Word of God has to say about it.

Christians, at least those grounded in God's Word, are told that *any* sexual activity outside marriage is wrong. As we are often asked, "Who are we to say what's right and wrong?" the question really devolves to the value one places on the Bible. Indeed, as pointed out in a previous chapter, the very concepts of right and wrong are called into question. The Bible states very clearly, either directly or indirectly through obvious inference, what is and is not permissible regarding most human behavior both between individuals and before God.

When we leave the guardrails of God's directions for living rightly, we are not just guilty because someone may make us feel that way; we are guilty in truth. Wrong is just that because God says so, and this verdict is not subject to opinion or the morphing of societal values. In other words, it is wrong whether you think so, your community thinks so, or the law says so. Now, of course, if there is no God, all this changes and right and wrong either don't exist at all (which is the fact of the matter) or they are indeed subject to what you think, your community thinks, or the law says. Again, without God you are loosed upon the earth to behave however you like, and so is everybody else.

The bottom line is that God's Word forbids homosexuality. The main point to focus on here is the *behavior*. How one comes to consider themselves homosexual is beside the point just the same as anyone is tempted to a tendency toward any misconduct. The question in all such cases is whether we *act* upon the temptation. *Considering* a wrong is temptation. *Performing* the act is sin. The world today attempts to whitewash this by mandating so-called gay marriage.

We should, however, never forget that as Christians we all called to love everyone. The behavior of an individual can be condemned, but we should always pray for them and provide whatever support we are able to help them in dealing with their internal conflict—whatever it may be. The arms of Christ are open as the arms of the church should be. We are all involved in the battle with control over our spirit, and the man battling pornography is fighting the temptation to act on his tendency as much as the homosexual.

Some want to claim that the Bible is silent regarding homosexuality. I'm not sure what book these folks are reading, but mine says:

> You shall not lie with a male as with a woman. It is an abomination. (Lev. 18:22)
>
> If a man lies with a male as he lies with a woman, both of them have committed an abomination. They shall surely be put to death. Their blood shall be upon them. (Lev. 20:13)
>
> Do you not know that the unrighteous will not inherit the kingdom of God? Do not be deceived. Neither fornicators, nor idolaters, nor adulterers, nor homosexuals,

nor sodomites, nor thieves, nor covetous, nor drunkards, nor revilers, nor extortioners will inherit the kingdom of God. (1 Cor. 6:9–10)

For this reason God gave them up to vile passions. For even their women exchanged the natural use for what is against nature. Likewise also the men, leaving the natural use of the woman, burned in their lust for one another, men with men committing what is shameful and receiving in themselves the penalty of their error which was due. (Rom. 1:26–27)

The Only Way

There are many who take exception with the biblical declaration that Christ is the only way to God the Father. As Paul points out in Romans, we are all separated from God. "As it is written: 'There is none righteous, no, not one; there is none who understands; there is none who seeks after God'" (Rom. 3:9b–11). In order to re-establish a relationship with him, some way to erase the sin debt attributed to us has to be found. Paul, John—in fact, all the apostles—preached Jesus as the doorway to God.

Initially, this One Way to enter heaven was a bitter pill to swallow. With the growth of the church, it eventually became the belief of the Western world. Christianity, with the Age of Exploration (1450–1750), became relegated to a relatively small part of the globe. As each new region was "discovered," there were religious beliefs revealed that were totally alien to Christianity. When Christ was claimed to be the sole way to salvation, it was seen as being narrow-minded and arrogant. Some wondered how this all-loving God could consign so many people to perdition.

Today, we have seen a shift toward the acceptance of all faiths leading ultimately to God. The Dogmatic Constitution of the Church was published by the Second Vatican Council, which calls for, among other things, all Roman Catholics to pray for those of another faith rather than *for* them to come to a saving faith in Jesus Christ. Sadly, some Protestants are adopting this position as well. They suggest that Christian missions should become a sort of Peace Corps rather than taking the gospel as their main purpose.

Then there are those who have adopted a pluralistic attitude, which, in one form, simply states that it is wrong to claim that one faith is true and all others are false. Pluralism states that *all* religions are true and lead to God. This stance is, well, just silly. If one faith is true, in fact, the others *cannot* be true. This would present a contradiction. They further state that the notion of the One Way is, once again, sheer arrogance. Granted, there are certainly arrogant Christians just as there are arrogant atheists, farmers, dentists, lawyers, carpenters, etc., which is indeed unfortunate. But that has nothing whatever to do with the validity of the message. If Christianity is true, why would we not want to share the good news with others? I would think that it would be the height of arrogance to keep it to ourselves!

Hell

Next to Jesus being the only way to heaven, the loudest complaint is about an all-loving, all-powerful God consigning so many people to eternal perdition. This is a complete misunderstanding of the premise. We consign *ourselves* to perdition by not accepting God's peace offering. But for those who still don't buy this explanation, I offer something else. We will look into not only the justice of God's consignment of sinners to perdition but also the fact that we leave him no other choice.

The truth around the justice of God's condemnation of sin can be seen by considering two things: man's sinfulness and God's sovereignty. As for man's sinfulness, it can be viewed as both infinitely evil and consistently measured out to all men.

If we think of how awful sin is when compared to the holiness of God, and that all men are guilty of it, then it follows that all men *deserve* to be punished; and by deserving punishment, God is *just* when he administers it. In fact, his righteousness demands it. To say that one deserves punishment and that it is not justly delivered is a contradiction.

Every crime has a greater or lesser punishment assigned to it based on the seriousness of the offence. Murder has a greater assignation of punishment than running a stop sign. While murder may carry the death penalty, the most severe sentence, it is not unjust because it is severe if it is proportionate to the crime. A death sentence for running a stop sign by comparison would be horribly unjust in anyone's mind.

A crime is more or less horrible according to the degree of obligation we are under. Societies laws dictate that we are under an obligation not to kill another citizen wrongfully, steal their property, or otherwise molest them in any way. The crime of hating another, if crime it is, is more or less grievous in proportion to our obligation to love that person. The degree of disobeying another—say, a parent or police officer—is more or less grievous depending on our obligation to obey them.

The obligation that we hold is based on the person to be obeyed's degree of loveliness, honorableness, or authority. When we say a person is lovely, contrary to common useage today, the words mean that they are worthy of being loved. As for honor, if one is more honorable than another, the more honorable places us in a more obligated position to hold them in honor. If one has more authority, we are more obligated to obey them.

In the case of God, he is infinitely lovely, infinitely deserving to be loved. He is infinitely great, glorious, and majestic so is infinitely honorable. And in authority, since he created everything, owns everything, and we are dependent on him in every way, he possesses infinite authority and is infinitely deserving of our obedience.

It follows that a crime against God, sin, puts us in infinitely grievous fault and deserving of infinite punishment. This is only common sense if we adhere to the premise that the punishment must be proportionate to the crime. Before God, there are no crimes of degree because of his holiness. *All* sin demeans his dignity.

> If one man sins against another, God will judge him. But
> if a man sins against the Lord, who will intercede for him?
> (1 Sam. 2:25)

This was why Joseph was so aghast at the idea of laying with Potiphar's wife.

> There is no one greater in this house that I, nor has he kept
> anything back from me but you, because you are his wife.
> How then can I do this great wickedness, and sin against
> God? (Gen. 39:9)

David feared to face God's judgment for the same reason.

> Against You, You only, have I sinned, and done this evil in
> Your sight. (Ps. 51:4)

Their fear was in facing the infinite punishment that they knew they would be justly facing because they understood who they would be sinning against. They knew this because any crime, if it was a crime at all, against an infinite being would carry such a sentence. While the infraction may be viewed as slight in man's eyes, it was not man who would stand in judgment. If we imagine a line of infinite length but possessing no width, it is nothing. But that same line, being only the thickness of a hair, exceeds any quantity of any finite thing. In like manner, a slight infraction to man has an infinite quantity to God.

Further, that God is just in inflicting eternal punishment can be seen by looking at just how much sin man is guilty of. Now, from what we've seen, we know that even one small sin is enough to make man eternally guilty. But man is full of sin, rotten with it. We are totally corrupt in every member of our bodies and all facets of our thoughts. Heads, hearts, bodies, senses are all totally depraved and only channels of corruption.

> For I know that in me (that is, in my flesh) nothing good
> dwells; for to will is present with me, but how to perform
> what is good I do not find. (Rom. 7:18)

Sinners are guilty of pride, maliciousness, contempt, blasphemy, and bickering with God. There are also obstinancy, perverseness, and hard-heartedness. All this and more are impervious to threats, promises, judgements or mercies, even the shed blood of Christ will not move the heart of an evil man.

And this is not taking into account the crimes of *commission*. Every command is broken, mercy and justice are despised (unless for yourself), and the honor of each person of the Trinity is blasphemed. If one sin is worthy of infinite punishment, how much more so when we look at all the sinfulness of man?

Alongside man's corruptness, if we view God's sovereignty, we will see how hollow are all man's objections of the harshness of the penalty. I don't claim to understand the full nature of God's sovereignty, but I do know that it includes the following:

God is under no obligation to keep man from sin, and he may choose to leave them to it. It is unreasonable to hold God responsible if he has created a reasonable creature capable of knowing God's will, receiving and understanding his law, and causing him to be aware of his moral pronouncements, and at the same time make it *impossible* for that same reasonable creature to sin or break his laws. If God were so obligated, it would make his provision of such laws and moral pronouncements meaningless. What need would there be for the threat of God's judgment let alone need of a Savior?

If viewed in the light of God being obligated to make man incapable of sin, then it is God who is under law. It now becomes his responsibiltiy and not the individuals to ensure that his law is obeyed. Some would say that since God permits man to sin, that man has God's permission to do so, and by this supposed permission, God has actually so ordered and arranged the event. This is the degree of absurdity that such a proposition leads us to.

God, in his sovereignty, has a right, once men sin, to determine whether or not he will provide a means of redemption. He may choose to redeem all, or some, or none at all just as He sees fit. He may take or leave whomever he so chooses. It is altogether fitting for God to so arrange things because of His supreme greatness. As creator of all things and owner of all things, he should make everything run according to his pleasure. His wisdom is infinitely superior, and there is no one who can add by counsel to his perfect decisions. As his creation, we are all completely dependent on him and can provide nothing to him.

How is what we've said to be applied?

As an unbeliever, a refuser of God's plan of redemption through Christ, you are under a just comdemnation. The punishment is fearsome, and you have cause to be afraid of it, but it is nonetheless just. If you believe otherwise, it is simply because you have remained insensible to what sin is and how guilty of it you are. To be sensible of it, you need only consult your conscience—look honestly at your past life. Consider the pattern of your life that you have maintained.

What have you done with all the days and nights God has granted you thus far? What have you done with the mercy of breath and strengh and God's daily provision?

How many bad things have you thought or actually done? How many times have you used God's name as a curse or that of his Son? Have you spent as much time in God's word as you have playing golf, watching sports, or other meaningless activities? How many immoral thoughts and acts have you committed? How many Sundays have you dismissed any thought of God and spent your time instead with equally corrupt friends? And are you raising your children to follow in your footsteps?

And how have you acted toward your parents? Have you honored them in the way you should? Have you criticized or even cursed them behind their backs or, worse yet, to their very faces? Have you obeyed them as you should recognizing their authority over you?

> The eye that mocks his father, and scorns obedience to his mother, the ravens of the valley will pick it out, and the young eagles eat it. (Prov. 30:17)

And how have you treated your neighbors? Have you hated others and been glad when some misfortune befell them? How many times have you been angry with someone over some slight to you or just because you were in a mood?

How much love have you expended on things of the world, desiring and even coveting something that you don't have? Have you ever envied someone their success or possessions? For the things of the world, have you ignored God and given even less consideration to Christ save in cursing? And how much time have you spent in worry about such things, expending your energies, attention to your family, and all your good intentions?

Do you possess pride in what you have been blessed with as if it were achieved by your efforts, shown off your money or talents or possessions? Who gave you that strength of mind and body that has allowed you to achieve? And how have you shouted and promoted disunity by insisting on your position?

How sensual and lascivious have you been, involved in pursuing desires that should not even be mentioned? How have you destroyed your family or another by seeking to accomplish your uncontrolled passions? How many others have you corrupted with your immoral actions?

> For it is shameful even to speak of those things which are done by them in secret. (Eph. 5:12)

And how some of you have lied beginning as children and carrying it into adulthood. What deceit and fraud have you committed in business or with you neighbors? How have you behaved among your family? What stupidity and drunkeness have you been guilty of and still shown no repentance or, if you did, were not sincere and reverted back to your old ways as soon as the heat was off of you?

Looking at all this, can you honestly consider God unjust if he does not show you mercy? Is he obligated to show you any? How can you consider, in light of his holiness compared to those things that you are guilty of, that he is harsh in his judgments? If you still think this, you cannot have any sense of your own guilt.

Next, let us consider how proportionate to the crime God's judgment is:

> With the merciful You will show Yourself merciful; with a blameless man You will show Yourself blameless; with the pure You will show Yourself pure; and with the devious You will show Yourself shrewd. Ps. 18:25–26

Regardless of how much you may consider this unfair, or that you may have come to fear it, if you were to receive such a judgment, it would only be what you deserve. God would deal with you exactly as you have dealt with him.

Let's consider the justness of God in his punishment based on how you have treated *him*, his Son *Jesus Christ*, your *neighbors*, and *yourself*:

God: You have never shown the least consideration toward God or shown any love for him. When a sinner comes to God, there is so much divine love expended as to be incomprehensible. This person has become aware of what they are and thrown themselves on God's mercy, repented, and been received into God's family. You have never seen any reason for repentance, so why should he be obligated to show you any mercy? Whether God is happy or not, worthy to be loved or not, means nothing to you. Why then should he be obligated to make you happy or show any love toward you? Why should God be considered harsh if he is unconcerned with your well-being?

Is it not true that as long as you were able to gratify yourself that you gave no thought to how it dishonored God? Why should God be forced or obligated to change your nature so that you can be less miserable than you are? And is it unjust that Christ, who you have only used as a curse, refuse to share his suffering and shed blood with you?

Maybe you have shed tears in apology for some of the things you've done and been ignored by God. Such people believe their souls are precious and deserve to be preserved in spite of themselves. God's holiness is far more precious that your unrepentant soul. You know when you do things that are wrong. God has constantly called you, exhorted you, and put others in your way that have begged you to change; and you have ignored it all. And yet God is harsh to ignore your occasional tear.

And you have not only ignored God in past times but also continue to do so. Your apologies have nothing to do with repentance but are only due to having been caught and brought to account. You still don't believe in Christ and have no love for God in your heart but are only sorry that you are in a bad way. Any honor or submission you show him is only a facade. It is only hypocrisy and a hope of escape.

And why shouldn't God cast you aside because you are also ungrateful for the mercies that he shows you on a daily basis? He has provided you with food, clothing, and preservation of life; and you have used them all to gratify yourself. You may have been given recovery from illness or protection from accident; but you have continued in your selfishness completely ungrateful except, perhaps, for the moment. You consider God obligated to continue these mercies.

You have decided on your way of doing things and stand not only ungrateful to God but also in very opposition to him. You have decided to stand against him in spite of his calls to you and the mercy shown. You have given yourself up to serving Satan, so why shouldn't God give you up to the side you've chosen? Why should God be obliged to rescue you when you have spent your life opposing him?

> Go and cry out to the gods which you have chosen; let them deliver you in your time of distress. (Judges 10:14)

Now, in honesty, have you not been encouraged to continue in your ways because you have been told that God is infinitely merciful and you can turn to him at the last moment? This is a most mistaken presumption on your part. Is God to be mocked by a mere lump of clay? Do you not know that God knows your heart and your deceit? Yet God is harsh to punish you eternally when on top of all your other slights to him, you attempt to connive and think that he can be duped. Really?

There is something especially base in trying to take advantage of a being possessed of infinite mercy and grace. This mercy, grace, and love should win your heart, not devolve into an even lower form of ingratitude and deceit. But instead of repenting, your are drawn deeper into sin:

> Or do you despise the riches of his goodness, forbearance, and longsuffering, not knowing that the goodness of God leads you to repentance? But in accordance with your hardness and your impenitent heart you are treasuring up for yourself wrath in the day of wrath and revelation of the righteous judgment of God. (Rom. 2:4–5)

Your ingratitude is so great, your self-love so entrenched, that you use God's love only as an encouragement to continue in your ways, counting on his rescue then your days come to an end.

Some of you have complained about how God has ordered things. God should have no objection to promiscuity, abortion, homosexuality, and so on. God should have so ordered the universe that there is no such thing as sin just as many of you would like to believe. Do you think that God must take your complaining and blasphemy as some sort of right that you possess? Do you think that because he allows you to do so that there will be no consequences?

> But indeed, O man, who are you to reply against God? Will the thing formed say to him who formed it, "Why have you made me like this?" (Rom. 9:20)
>
> Woe to him who strives with his Maker! Let the potsherd strive with the potsherds of the earth! Shall the clay say to him who forms it, "What are you making?" (Is. 45:9)

Jesus: Consider how you have treated Jesus Christ. God would be just in light of your great sin to eternally punish you, but he hasn't done that. He has provided a means for your redepmtion through his Son, the only Savior of men. Any man or woman that is not eternally punished is saved only by him, and we have God's promise that if we come to him, we will not only not suffer eternal punishment but will enjoy eternal life with him.

But do you receive Christ with gratitude for all he suffered so that you might be saved? No, you do not. There is a big difference between your willingness not to be eternally punished and your receiving Christ as your Lord and Savior. No doubt you are willing to escape misery. You speak of a willingness to avoid punishment but are still unwilling to think that God should have any say in the way you live. In other words, such a proclamation on your part is only further evidence of your undying selfishness.

Coercion and freedom cannot coexist. You feel you are forced to give lip service to Christ in order to avoid punishment. This will not do as Christ still has no place in your heart; free and willing acceptance is the only condition acceptable. He doesn't want you to come grudgingly or against your will. Is it reasonable to think that Christ will force himself on you?

Futher, how can it be possible that you would accept Christ as your Savior from eternal punishment when you are not convinced that you are deserving of such a fate? Indeed, if you think that you are not deserving of such, the offer of Christ as a Savior can only appear to you as an insult. If you are not conscious of any guilt, then the very offer of Christ's satisfying that guilt by taking it upon himself would imply that you are, in fact, guilty. It is impossible that you would be willing to accept such an offer if you are unconvinced. To act as if you accept it while all the time not believing in your guilt is but another facade.

You cannot accept it further because you have not recognized God's sovereignty and his right to condemn you. How then can you accept atonement from one you think unjustly condemns you?

In addition, you cannot accept Christ as Savior because you cannot admit that your own goodness is worth nothing. This is evident because you make so much of how good you are both publicly and in your own heart. You think how much more you have done than others, how much more you have contributed, how high you stand in your community. Surely God should have some reward in store for you for all the great and wonderfull things you have done.

With this in mind, how can you be grateful to a God who makes nothing of all your works and accomplishments? If you have so much pride in your own greatness, how can you place yourself in the position of a lowly worm in need of a Savior?

Seeing that you have never freely seen the need to be saved and have thereby rejected God's provision of a Savior, how is it unjust that Christ not save you? Not only is it just, but it is manifestly just because mercy itself has been rejected. What more can God do to uphold his holiness that to offer his mercy with the sole proviso that you repent and sincerely accept Christ? And this being rejected will any have the gall to say the sentence is unjust?

God will surely be even more aggreived with you. In addition to your sinfulness, you reject his offer of forgiveness! Jesus, who freely and lovingly knocks on your door, is sent away!

Others: Let us next consider how you treat others. Have you not laughed and poked fun at your friends and acquaintenances who have sincerely repented of their past and sought through Christ to change their lives? In fact, you may have come to avoid them and their converstion especially when such a one, who knows you, counsels you on your behavior. And your knowing them from the past have made fun of them and doubted their sincerity trying with all your might to pull them back.

You have been against any who comes to accept the deliverance you have rejected. If you have so little love for your friends that you would begrudge them their newfound joy, is that not further evidence of your stand in outright opposition to God? Not to mention your true feelings toward your friends and neighbors.

Not only have you begrudged others their salvation but you have assisted others in deserving eternal punishment. You have assisted others in the corruption of their minds and in commission of sins before both God and man. Indeed, you have done so to your own children by neglecting their education, not leading them in the right ways of going and by your bad example. You have left them adrift to find their own way, and given the condition of the world, where do you think they will end up without some divine intervention? And will they receive that intervention because of any thanks to you?

Self: Lastly, consider how you treat yourself. You have not given a rip about your own salvation. God has called, counseled, demanded, and threatened; and still you haven't paid any heed. Is God obligated to care more for your welfare than you do? Should God be expected to take care of you when you are determined to destroy yourself? How is God unjust if he allows you to continue on to the reward you have chosen for yourself?

Not only have you neglected your salvation but also have actively been at work to destroy yourself. You have been adamant in your ways in spite of being in possession of the gospel and having an abundance of pastors and churches available to help you along the way. You may even had father, mother, brother, sister, friends all begging you to listen but have paid them no mind. You have willfully and consciously destroyed yourself despite God's earnest desire otherwise. Since you have decided your path, who else is to blame if you are punished? It is only just of God to let you have your own way.

We have considered these things because unless you are senseless or irretrievably perverse, you will understand that you stand before God justly condemned.

In spite of all we have said, though it would be right for God to eternally condemn you, it would also be right for God to save you through Christ. God can, through Christ, justly and with honor to his holiness show you mercy even now. That is how precious the shed blood of Jesus is to his Father. The sacrifice of Christ was a much greater deed than that of throwing all of us into perdition. Christ is so great in the eyes of God that even though Christ suffered at the hands of sinners, God is willing to be at peace with us who accept that sacrifice regardless of our unworthiness.

It is only by this that we are saved because all our works, prayers, and church attendance are nothing compared to the actions of Christ. So there is an encouraging word for you. The door may still be open to you, and God, in his infinite love and mercy, may yet let you in.

For those of you who are Christians, be not proud because you were of the same cloth. It would have been just if God had eternally consigned you to punishment. It is through no goodness of your own that you have been rescued. You too did not love God, yet he has still shown such love for you as to be beyond understanding. Are you not ashamed for not praising God twenty-four hours a day?

It is God and God alone who has opened your eyes and hearts. I know because I too was such a one.

(Jonathan Edwards gives this topic a thorough treatment in a sermon he delivered around 1743 during the Great Awakening. I highly recommend his works as, in my opinion, he was the greatest theological mind this country has produced.)

What about Church Attendance?

The church is not a building. Everyone who has accepted the truth of Jesus Christ is already a member of his church. It is the body of believers—all believers in Jesus. Now, of course, just what you believe in matters and we have outlined the criteria in preceding chapters. As a body of believers, we are segmented into various denominations, lamentably, and they meet in some sort of building ranging from a barn to a megachurch. Being a part of one of these is important to your individual growth in knowledge of God, and by being among kindred spirits, you will find comfort, support, and fellowship. Jesus told us that "wherever two or more are gathered in My name, there I will be." If for no other reason you should want to be where he is. Besides, if you have truly come to have some understanding of God's love for you and his free gift of reconciliation, Jesus, you should look forward to an opportunity to worship and praise him.

Now, the tough issue: "Which group of believers should I worship with?" I'm going to upset some people with this, but I frankly don't give two cents for denominational identifications. I have attended biblically sound churches of nearly all of them. What is important is that Jesus Christ is preached as the salvation of man, and this can only be done truly if it includes all of scripture, not cutting out the inconvenient parts. God needs to be shown as the Creator of all things—in six days. The Bible needs to be presented as what it is, history—*all* of it. If a pastor has a problem with any of this, look somewhere else.

Attending a local church with friends can be a wonderful thing but only if the Word of God is being preached, and believe me, not every church is doing that. I personally fear that fewer and fewer are. It will not help you to attend because of the glitzy youth programs, the band that leads the music, or all the picnics and special events that are held, if Jesus, and him crucified and risen again in payment for *your* sin, is not the focus and motivation for being there. And don't be lured by cults like the Church of Jesus Christ of Latter-Day Saints (Mormons) and Jehovah's Witnesses. They have absolutely nothing to do with the good news.

Epilogue

And so we have reached the end of our journey. In my case, at least, it was the end of one and the beginning of another. The journey that I am on now is the *real* trip of my life. Prior to my acceptance of those things that we have looked at together, I was a ship without a tiller. The path that we have covered took me on many winding roads all but one of which proved to be dead ends.

During this time, my relationships with others were ill defined and my understanding of myself even fuzzier. To paraphrase the Ghost of Christmas Past in the movie *Scrooged*, "I didn't know who I was, what I wanted, or what the hell was going on." There was just something wrong, something missing, or maybe it was just something that I had missed. Other people didn't seem as aimless as I. Why in the world had I been put here? What purpose did I serve? What purpose did *anyone* serve?

Along the way, I had the opportunity to get to know many wonderful people (and some who were definitely not so wonderful), but lasting relationships didn't seem to be something I was able to adjust to. I just didn't know what was going on! But *something*, about thirty-five years or so ago, set me off in a specific direction. I believe it may have been the first *purpose* in my life. I had to *know* what that missing piece was. In any event, that purpose led me to the conclusions that I have drawn as delineated in this book.

Evolution, humanism, materialism, or whatever you choose to call it is nothing more than a false religion. To give the followers of such due credit, if credit it can be called, they knew the correct methodology to follow in order to get their lies accepted by the general public. They are the same principles used by all subversive elements that seek to topple freedom. Like the many socialist groups in Russia prior to the 1917 revolution, they realize that it was enough to infiltrate the media and universities and bide their time. Let me offer a few quotes and see if you can guess who said them:

> To attract people, to win over people to that which I have realized as being true, that is called propaganda. Propaganda should be popular, not intellectually pleasing. It is not the task of propaganda to discover what is intellectually true.
>
> The best propaganda is that which, as it were, works invisibly, penetrates the whole of life without the public having any knowledge of the propagandistic initiative.
>
> The principle and which is quite true in itself and that in the big lie there is always a certain force of credibility; because the broad masses of a nation are always more easily corrupted in the deeper strata of their emotional nature than consciously or voluntarily. . . . The bigger the lie, the more it will be believed.

The first quote was delivered by Joseph Goebbels to Nazi party members on January 9, 1928. The next was provided to posterity by the same man in March 1933 and the last, Goebbels again, in Munich on January 12, 1941. If you are too young to know who Joseph Goebbels was, look him up. It will provide you with a real education on just how warped and depraved a man can be.

The usefulness of the media is expressed in the following:

> The press should be not only a collective propagandist and a collective agitator, but also a collective organizer of the masses.

> Our program necessarily includes the propaganda of atheism.

> He who now talks about the 'freedom of the press' goes backward, and halts our headlong course towards Socialism.

We can thank Vladimir Lenin for these encouraging words. He was the instigator of the most insidious regime the world has ever seen. He and his successor, Joseph Stalin, imprisoned, impoverished, and murdered tens of millions of their own people.

I am not saying that all atheists are Nazis or communists, but it is more than coincidence that Nazis and communists look to the same sources for their justification. There is no example that I have been able to find where a follower of Jesus Christ was either.

> We must, however, acknowledge, as it seems to me, that man with all his noble qualities . . . still bears in his bodily frame the indelible stamp of his lowly origin.

> I love fools' experiments. I am always making them.

> Man is descended from a hairy, tailed quadruped, probably arboreal in its habits.

> False facts are highly injurious to the progress of science, for they often endure long; but false views, if supported by some evidence, so little harm, for everyone takes a salutary pleasure in proving them false.

Courtesy of Charles Darwin perpetrator of the second greatest lie ever told. The first was "Surely you shall not die."

> Biology is the study of complicated things that have the appearance of having been designed with a purpose.

> Religious fanatics want people to switch off their own minds, ignore the evidence, and blindly follow a holy book based upon private "revelation."
>
> Design can never be an ultimate explanation for anything. It can only be a proximate explanation. A plane or a car is explained by a designer but that's because the designer himself, the engineer, is explained by natural selection.
>
> What has "theology" ever said that is of the smallest use to anybody? When has "theology" ever said anything that is demonstrably true and is not obvious? What makes you think that "theology" is a subject at all?
>
> Richard Dawkins, current godfather of the church of materialism and funniest man I have ever listened to.

I am personally happy that Dawkins is an outspoken mouthpiece for materialism because there is little that I find more hilarious than his books and interviews. It is curious to me that so many evolutionists today are concerned over their lack of progress in proving their case and yet persist in it. It is known by those who view the evidence (laughably, Dawkins professes to be one of these) that current discoveries are leading us further away from Darwinism. Granted, most die-hard materialists are still unwilling to accept God as an answer, and many others will only go halfway espousing intelligent design (ID).

There are several reasons for their persistence, and they are as old as my giddy aunt. One is simply a matter of money. If these learned men and women were to say that their next work was going to be in using the evidence of molecular biology to prove the existence of design, they would lose their grant money and quite possibly their employment. That is not an acceptable field of study in academia today. Another reason comes from the equally unacceptable (at least as far as the search for truth is concerned) matter of human pride. Any man or woman that has spent decades promoting a lie would be a very unusual person indeed to admit to being such a promoter now. There is another reason that some claim, and that is that they are intellectually unable to accept God really exists. What this really means is that they have some part of their lives that they feel they will be forced to give up if there is a god that they have obligations toward. There are many who would be loath to give up their participation in the destructive behavior that has all but taken over our society. It is hard enough for a drunkard to admit to being one, let alone being told that his/her behavior is wrong. The same with the gay person, the promiscuous person, the unethical person, the liar, the adulterer, the hateful, the cruel, and on and on. It is curious that these folks who claim that they want "freedom" to do as they please are in reality slaves to their passions, lacking any self-control, self-respect, or respect for others. But freedom to choose is theirs indeed.

It is true, as loudly proclaimed by many, that God cannot be proven to exist through empirical experiment any more than evolution. It is also true that Jesus Christ, therefore, cannot be proven to be his Son, to have risen from the grave, or to have been an atonement for our sin. What we can prove is that the many of the ideas contained within the counteroptions available to us *can* be demonstrably proven to be *unworkable*. While I have only made a cursory examination of evolution in this book, there are many other authoritative books available, with more coming practically daily, that I invite you to read. Only a foolish man, convinced of his own wisdom, continues on a road known to be wrong.

Materialists strike me as being locked in a very large room. In this room, they believe, is everything that can or will be known. They keep digging and scratching and theorizing, convinced that if they just dig deeper or scratch harder or theorize more confidently that they will find what they are looking for. All the while, in one of the walls, there is a door but the diggers cannot see it. Not only can they not see it but they also refuse to acknowledge its existence. Occasionally one of the diggers will break away from the group, see that the door does exist and walk over to it. They even have the courage to open it. But they will not walk through into the light of the amazing world that it reveals to their eyes. These are no longer diggers but have become intelligent designers. These are neither the hot or cold but those whom God vomits out of his mouth.

Charles Darwin said himself that "to kill an error is as good a service as, and sometimes better than, the establishing of a new truth or fact."

The alternative view, the God of the Bible, has been available to us for thousands of years. We who have come to believe in its trustworthiness are the ones who are free. God does not give us "thou shalt nots" to ruin our fun but in order that we might live peaceful, fulfilled lives because our Creator knows what will make us peaceful and fulfilled. If your car broke down (I should say when), would you prefer to have it looked at by the engineer who designed it or Bobby at Billy Bob's Shade Tree Auto Service? Our designer, our Creator, our God knows how to fix us better than any Richard Dawkins of "the Acme School for Unintelligent Noncauses." He provides us with guardrails. He provides us with reason and intelligence, and he rejoices when we discover new pieces of evidence of his glory in science.

Jesus did not ask his disciples how they *felt* about things. Sooner or later, important decisions made on the basis of your feelings can, and probably will, lead to disaster. We were given a brain to think, and it is with this particular organ that we should formulate our decisions. No, Jesus asked his disciples what they thought:

> When Jesus came into the region of Caesarea Philippi, He asked his disciples, saying, "Who do men say that I, the Son of Man, am?" So they said, "Some say John the Baptist, some Elijah, and others Jeremiah or one of the prophets." He said to them, "But who do you say that I am?" Simon Peter answered and said, "You are the Christ, the Son of the living God." Jesus answered, "Blessed are you, Simon Bar-Jonah, for flesh and blood has not revealed this to you, but My Father who is in heaven." (Matt. 16:13–17)

Jesus desired his disciples to look at the evidence and then to decide their answers. Some merely relayed hearsay and did not answer the question with a direct response. Peter, bless his heart, told him what had been revealed to him. Which brings us to another point. Why do many view the Bible as they do, just a book made up by a bunch of confused Jews? Just like Peter, spiritual truth can only be understood by those who have been spiritually awakened, and this comes to those who accept Christ and are indwelled by the Holy Spirit. With the aid of the "comforter," the more you read the Word of God, the more it is revealed to you as such.

I invite you to try it out for yourself. We are told that if we sincerely call on him with open hearts that God's Word will not return void. Consider what great men have witnessed to Christ with their lives. Consider the thousands of martyrs both past and present who have testified to Christ with *their* lives. Consider how ardently people continue in the truth today. This is not to say that Christians are better than others. We still possess our fallen nature and will until we pass over into heaven or Christ returns. There are hypocrites who use the name of Christ for their own selfish reasons, and they will get their just reward; God will not be mocked. There are many who claim to be Christians that don't really know what it means to profess that name. But this need not be you. Walk through that door trusting that the world it reveals is the true one and see what mercies and wonders it contains for you.

> At that time the disciples came to Jesus, saying, "Who then is greatest in the kingdom of heaven?" Then Jesus called a little child to Him, set him in the midst of them, and said, "Assuredly, I say to you, unless you are converted and become as little children, you will by no means enter the kingdom of heaven. Therefore whoever humbles himself as this little child is the greatest in the kingdom of heaven. Whoever receives one little child like this in My name receives Me." (Matt. 18:3–5 NKJV)

Jesus calls us to accept the evidence, trusting as little children in its truth. He does not ask us to check our brains at the door. But when all is said and done, we must trust in things that can't be seen. In this way, we are like materialists though in their case, they are trusting in what is *worthless*. They are trusting in a world view based on error, deceit, and willful misrepresentation of fact.

> Now faith is the substance of things hoped for, the evidence of things not seen. For by it the elders obtained a good testimony. By faith we understand that the worlds were framed by the word of God, so that the things which are seen were not made of things which are visible.

The hope spoken of here is the hope of assurance and not wishful thinking. As with children hoping that Christmas will come, this is not a false hope because it is certain that day will surely come. Our faith is that of promises—promises received from One who cannot lie.

> Then Jesus said to those Jews who believed Him, "If you abide in My word, you are my disciples indeed. And you shall know the truth, and the truth will make you free." (Jn. 8:31–32 NKJV)

For those of you that have been bold enough (or curious enough) to have followed me thus far, you may have found that you don't want this journey to end. It doesn't have to, in fact, it can just be the beginning of a new, far more satisfying one. One that you can proceed on with clarity. You don't need to pack or make reservations.

If you accept my conclusions, you merely need to admit your need of a savior and recognize that that Savior is Jesus Christ. Open your heart to him in sincerity and repentance and he will enter. You are no worse (or better) than any other person in God's eyes. And those of us already in his family will welcome you with open arms. I pray that you will seriously consider your world view. Trust me when I say that the view is far more pleasing and easier to live with when you can see clearly.

> Behold, I stand at the door and knock. If anyone hears My voice and opens the door, I will come in to him and dine with him and he with Me. (Rev. 3:20 NKJV)

> Now to Him who is able to keep you from stumbling, and to present you faultless before the presence of His glory with exceeding joy, to God our Savior, who alone is wise, be glory and majesty, dominion and power, both now and ever more. Amen. (Jude 24–25)

Acknowledgments

I owe much to men that I have never known and will probably never meet this side of heaven simply because they are not among the living any longer. Some of them for quite some time. While it may seem strange, my path was eventually directed largely by the apostles Paul and John. Paul was frank and unsparing in his convincing me that I am what I really am, and John in convincing me that Jesus is what he is. Of course, the words of Christ himself are beyond comparison. Indeed, it is when we move further from these original teachings and confuse them with human ideas and explanations that Christians and people in general are more likely to run into confusion and error.

I would particularly like to thank two men because my remarks about modern philosophers were somewhat disparaging. William Lane Craig and Norman Geisler write in such a manner as to make their subjects both understandable and interesting. Things are really not as difficult as materialists would like you to believe, at least when considering the basics. Even when Craig ventures into time, space, and eternity, I understand him though I would caution the weak of heart to avoid those particular works. (Just kidding, sort of.)

I would also like to thank Tammy Jarvis, Jesse Bayne, Jessica Mowles, Michelle Salisbury, Darius Blevins, Trevor Durfee, Jesse Allen, Diane Rameriz, Amy Morgan, Barbara Steele, Sherrie Witt, Lenard Cook, Chris Riquelmy and Aelish Truver. Their understanding and consideration were a huge help in completing this project.

And lastly I want to thank Jean, Gabe, Eli and Tanner. It was with a heart full of love for you that compelled me to write this book. Your support through it all was only a further example of what makes us what we are.....family.

Sources and Suggested Reading

Geisler, Norman L & Turek, Frank, *I Don't Have Enough Faith to be an Atheist*, Wheaton: Crossway, 2004.

Craig, William L., Resonable Faith, Wheaton: Crossway, 2008.

Muncaster, Ralph O., Dismantling Evolution, Eugene: Harvest House, 2003.

Martin, Jobe, The Evolution of a Creationist, Rockwell: Biblical Discipleship, 2004.

Gish, Duane T., Evolution: The Fossils Still Say No!, E Cajon: Institute for Creation Research, 2006.

Gills, James P., Darwinism under the Microscope, Lake Mary: Charisma House, 2002.

Patterson, Roger, Evolution Exposed: Earth Science, Hebron: answersingenesis, 2008.

Patterson, Roger, Evolution Exposed: Biology, Hebron: answersingenesis, 2009.

McDowell, Josh, Evidence for Christianity, Nashville: Thomas Nelson, 2006.

Linder, William L. Jr., John Calvin, Bethany House, Amazon ebook, 2011.

Farrell, Vance, Making Archeology Biblical, Altamont: Evolution Facts, 2009.

Appendix A

First Dynasty		
Pharaoh	A.K.A	Approx. Dates (B.C.)
Narmer	Menes	3100–3050
Nor-Aha	Ity	3050–3049
Djer	Athothis	3049–3008
Djet	Zet	3008–2975
Merneith (Queen)		3008
Den	Udimu	2975–2935
Anedjib	Adjib	2935–2925
Smerkhet	Iry	2925–2916
Qa'a	Kaa	2916–2890
Second Dynasty (Sometimes combined with the first dynasty)		
		Date not known approx. years reigned
Hotepsekhemwy	Bedjau	38
Nebra	Raneb	14

Nynetjer		40
Senedj		20
Seth-Peribsen	Kaires	?
Sekhemib-Perenmaat	Sesokhris	?
Khasekhemwy		18
Old Kingdom (3rd - 6th Dyansties) **Third Dynasty**		
Djoser	Netjerikhet	19
Djoserty	Sekhemkhet	6
Nebka	Sanakht	9
Teti	Khaba	6
Huni	Qahedjet	24
Fourth Dynasty		
Sneferu	Nebmaat	2613 - 2589
Khufu	Medjedu	2589 - 2566
Djedefre	Kheper	2566 - 2558 ?
Setka?	Nebkare	2558 ?
Khafre	Userib	2558 - 2532
Menkaure	Kakhet	2532 - 2503 ?
Shepsiskaf	Shepseskhet	2503 - 2499 ?
Djedefptah		2499 - 2494 ?
Fifth Dynasty		
Userkaf	Irimaat	2494–2487
Sahure	Nebkhau	2487–2475
Neferirkare Kakai		2475–2455
Shepseskari Isi		2455–2448
Neferefre	Neferkhau	2448–2445
Nyuserre Ini		2445–2421
Menkauhor Kaiu		2421–2414
Djedkare Isesi		2414 - 2375
Unas	Wadjtawy	2375 - 2345

	Sixth Dynasty	
Teti	Seheteptawy	2345–2333
Userkare		2333–2331
Pepi I	Merenre	2331–2287
Nemtyemasf II	Merenre	2287–2278
Pepi II	Neferkare	2278–2184
Nemtyemasf II	Merenre	2184
Neitiqerty Siptah?	Nitiqret	2184–2181
	First Imtermediate Period (7th - 10th Dynasties) **Seventh Dynasty**	
	Netjerkare?	?
	Menkare	?
	Neferkare II	?
Nebi	Neferkare	?
	Djedkare	?
	Neferkare	?
	Merenhor	?
Seneferka	Neferkamin	?
	Nikare	?
	Neferkare	?
	Neferkahor	?
	Eighth Dynasty	
		Years Reigned
	Neferkare	?
	Neferkamin Anu	?
Iby	Qakare Ibi	2
	Neferkaure II	4
Khwiwihepu	Neferkauhor	2
	Neferirkare	1.5
	Ninth Dynasty	
Wakjare Khety I		?
Merykare I		?

Neferkare III		?
Wankhare Khety II		?
Tenth Dynasty		
Akhtoy Wahkare		?
Akhtoy Wahkare		?
Akhtoy Nebkaure		?
Merykare		?
Middle Kingdom (Eleventh - Fourteenth Dynasties) **Eleventh Dynasty**		
Mentuhotep I		2134–?
Intef I	Sehertawy	?–2118
Intef II	Wahankh	2118–2069
Intef III	Nakhtnebtepnefer	2069–2061
Mentuhotep II	Nebhetepre	2061–2010
Mentuhotep III	Sankhkare	2010–1998
Mentuhotep IV	Nebtawyre	1998–1991
Twelfth Dynasty		
Amenemhat I	Sehetepibre	1991–1962
Sesostris I	Kheperkare	1971– 1926
Amenemhat II	Nebkhaure	1929–1895
Sesostris II	Khakheperre	1897–1878
Sesostris III	Khakaure	1878– 1839
Amenemhat III	Nimaatre	1860–1814
Amenemhat IV	Maakherure	1816 - 1806
Sobekneferu (Queen)	Sobekkare	1806 - 1802
Thirteenth Dynasty		
Sobekhotep		?
Amanemhat V	Sekhemkare	?
Ameny Qemau		?
Qemau	Hotepibre	?
Amenemhat VI	Sankhibre	?

Nebnuni	Semenkare	?
Sobekotep II	Sekhemre-Khutaway	?
Renseneb		?
Hor	Auibre	?
Amenemhat VII	Sedjefakare	?
Wegaf	Khutawyre	?
Khendjer	Userkare	?
Intef	Shetepkare	?
Neferhotep I	Khasekhemre	?
Sobekhotep IV	Khaneferre	?
Sobekhotep V	Merhotepre	?
Sobekhotep VI	Khahotepre	?
Ibiaw	Wahibre	?
Ay	Merneferre	?
Ini I	Merhotepre	?
Sewadjtu	Sankhenre	?
Fourteenth Dynasty		
Nehesy		?
Sekheperenre		?
Merdjefare		?
Sekhaenre Yakbim		?
Qareh	Qar	?
Ammu		?
Maaibre Sheshi		?
Aperanat		?
Samuqenu		?
Meruserre Yaqub-Har		?
Second Intermediate Period (Fifteenth - Seventeenth Dynasties) **Fifteenth Dynasty**		
Salitis		?
Sakir-Har		?
Khyan		?

Apophis		1600 ?
Khamudi		1550 ?
Sixteenth Dynasty		
Djehuti	Sekhemre-sementawi	1649
Sobekhotep VIII	Sekhemre-seusertawi	1645
Neferhotep III	Sekhemre-seankhtawi	1629
Mentuhotep VI	Seankhenre	1628
Nebiryraw I	Sweadjenre	1627
Nebiriau II	Neferkare	1601
	Semenre	1601
Bebi-Ankh	Seuserenre	1600
Shedwast	Sekhemre	1588
Dedumose I	Djedhetepre	?
Dedumose II	Djedneferre	?
Montemsaf	Djedankhre	?
Mentuhotep VI	Merankhre	?
New Kingdom (Eighteenth - Twentieth Dynasties))		
Eighteenth Dynasty		
Ahmose I	Nebpehtire	1549–1524
Amenhotep I	Djeserkare	1524–1503
Thurmose I	Akheperkare	1503–1493
Thutmose II	Akheperenre	1493–1479
Hatshepsut	Maatkare	1479–1458
Thutmose III	Menkheperenre	1479–1424
Amenhotep II	Akheperure	1424–1398
Thurmose IV	Menkheperure	1398–1388
Amenhotep III	Nebmaatre	1388–1350
Akhenaten	Neferkepherure	1351–1334
Semenkhare	Ankhkheperure	1335–1333
Neferneferuaten	Meriwaenre	1335–1333
Tutankhamun	Nebkheperure	1333–1323
Ay	Kheperkheperure	1323–1319
Horemheb	Djeserkheperure	1319–1292

Nineteenth Dynasty		
Ramesses I	Menpehtire	1298–1296
Seti I	Menmaetre	1296–1279
Ramesses II	Usermaatre	1279–1212
Merneptah	Banenre	1212–1201
Seti II	Userkheperure	1201–1195
Amenmesse	Memire-Setepenre	1200–1196
Siptah	Sekhaenre/Akheperre	1195–1189
Twosret (Queen)	Sitre-Merenamun	1189–1187
Twentieth Dynasty		
Setnakhte	Userkhaure	1187–1185
Ramesses III	Usermaatre-Meryamun	1185–1153
Ramesses IV	Heqamaatre	1153–1146
Ramesses V	Amenhirkhepeshef I	1146–1141
Ramesses VI	Amenhirkhepeshef II	1141–1133
Ramesses VI	Itamun	1133–1125
Ramesses VIII	Sethhirkhepseshef	1125–1123
Ramesses IX	Khaemwaset I	1123–1104
Ramesses X	Amenhirkhepeshef III	1104 - 1094
Ramesses XI	Khaemwaset II	1094 - 1064
Third Intermediate Period (XXIst - XXVth Dynasties) Twenty-First Dynasty		
Smendes	Nesbanebdjed I	1077–1051
Amenemnisu		1051–1047
Pinedjem I		1062– 1039
Psusennes	Pasebkhanut I	1047–1001
Amenemope		1001– 992
Osorkon the Elder		992–986
Siamun		986–967
Psusennes II	Pasebkhanut II	967–943

Twenty-Second Dynasty		
Shoshenq I		943–922
Osorkon I		922–887
Shoshenq II		887–885
Takelot I		885–872
Osorkon II		872–837
Shoshenq III		837–798
Shoshenq IV		798–785
Pami		785–778
Shoshenq V		778–740
Pedubast II		740–730
Osorkon IV		730–716
Twenty-Third Dynasty		
Harsiese A		880–860
Takelot II		840–815
Pedubast I		829–804
Iuput I		829–804
Shosshenq VI		804–798
Osorkon III		798–769
Takelot III		774–769
Rudamun		759–739
Ini		739–734
Peftjauwybast		734–724
Twenty - Fourth Dynasty		
Tefnakhte I		732–725
Bakenranef	Bocchoris	725–720
Twenty - Fifth Dynasty		
Kashta		760–752
Piye		752–721
Shabaka		721–707

Shebitku		707–690
Taharqa		690–664
Tantamani		664–656
Late Period (XXVIth - XXXIst Dynasties) **Twenty-Sixth Dynasty**		
Psamtik I		664–610
Necho II		610–595
Psamtik II		595–589
Apries		589–570
Amasis II		570–526
Psamtik III		526–525

www.ingramcontent.com/pod-product-compliance
Lightning Source LLC
Chambersburg PA
CBHW020737180526
45163CB00001B/264